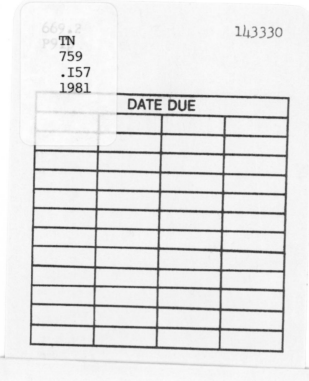

DATE DUE

Precious Metals

Proceedings of the Fourth International Precious Metals
Institute Conference, held in Toronto in June 1980

PERGAMON TITLES OF RELATED INTEREST

BOOKS

Ahrens	Origin and Distribution of the Elements, 2nd Symposium
Atkins & Lowe	The Economics of Pollution Control in the Non-Ferrous Metals Industry
Bailar	Comprehensive Inorganic Chemistry, Volume 3 — Groups IB, IIB, IIIA, VA, VIA, VIIA, VIII
Barrett & Massalski	Structure of Metals, 3/e
Biswas & Davenport	Extractive Metallurgy of Copper, 2/e
Coudurier et al.	Fundamentals of Metallurgical Processes
Gilchrist	Extraction Metallurgy, 2/e
IUPAC	Solubility Data Series
	Volume 3 — Silver Azide, Cyanide, Cyanamides, Cyanate, Selenocyanate and Theocyanate
	Volume 7 — Silver Halides
	Volume 12 — Metals in Mercury
Jeffery & Hutchinson	Chemical Methods of Rock Analysis, 3/e
Kubaschewski & Alcock	Metallurgical Thermochemistry, 5/e
Moruzzi et al.	Calculated Electronic Properties of Metals
Parker	An Introduction to Chemical Metallurgy, 2/e (SI units)
Savitsky et al.	Physical Metallurgy of Platinum Metals
Strasheim	Analytical Chemistry in the Exploration, Mining and Processing of Materials
Wills	Mineral Processing Technology

JOURNALS

Acta Metallurgica
Canadian Metallurgical Quarterly
The Journal of Physics and Chemistry of Solids
Materials Research Bulletin
Metals Forum
The Physics of Metals and Metallography
Progress in Materials Science
Scripta Metallurgica

Precious Metals

Edited by:
R. O. McGachie
and
A. G. Bradley

Pergamon Press
Toronto • Oxford • New York • Sydney • Paris • Frankfurt

Pergamon Press Offices:

Canada	Pergamon Press Canada Ltd., Suite 104, 150 Consumers Road, Willowdale, Ontario, Canada M2J 1P9
U.K.	Pergamon Press Ltd., Headington Hill Hall, Oxford, OX3 0BW, England
U.S.A.	Pergamon Press Inc., Maxwell House, Fairview Park, Elmsford, New York 10523, U.S.A.
Australia	Pergamon Press (Aust.) Pty. Ltd., P.O. Box 544, Potts Point, N.S.W. 2011, Australia
France	Pergamon Press SARL, 24 rue des Ecoles, 75240 Paris, Cedex 05, France
Federal Republic of Germany	Pergamon Press GmbH, Hammerweg 6, Postfach 1305, 6242 Kronberg-Taunus, Federal Republic of Germany

TN
759
,I 57
1981

14 3330

mu 1988

Canadian Cataloguing in Publication Data

International Precious Metals Institute. Conference
(4th: 1980: Toronto, Ont.)

  Precious metals

Bibliography: p.
Includes index.
ISBN 0-08-025369-5 (bound).

1. Precious metals — Congresses. I. McGachie,
R. O. (Richard O.), 1932 - II. Bradley, A. G.
(Anne Grace), 1947- III. Title.

TN759.I57   669'.2   C81-094222-4

In order to make this volume available as economically and as rapidly as possible, the authors' typescripts have been reproduced in their original forms. This method unfortunately has its typographical limitations but it is hoped that they in no way distract the reader.

Printed and bound in Canada by
T. H. Best Printing Company Limited, Don Mills, Ontario

Contents

International Precious Metals Institute   ix

Foreword   xi

Forty Years of Platinum: Some Recollections and Reflections
   Leslie B. Hunt   1

Mining and Extraction

Placer Mining in Alaska
   E. Beistline   9
An Overview of Gold and Silver Mining in Canada
   M. R. Brown   17
Mineralogy and Distribution of the Platinum Group in Mill Samples from
   the Cu-Ni Deposits of the Sudbury, Ontario, Area
      Louis J. Cabri   23
Silver Mining in Peru
   G. Florez Pinedo   35
Treatment of Complex Silver Arsenide Concentrate in Nitric Acid System
   W. Kunda   39
Geology and Tonnage-Grade Relationships of Bulk Minable Precious
   Metal Deposits in North American Cordillera
      Stanley W. Ivosevic   59

Recovery and Refining

Mechanical Processing of Electronic Scrap to Recover Precious-Metal-
   Bearing Concentrates
      Fred Ambrose and B. W. Dunning, Jr.   67
A Printed Wiring Board Manufacturer Looks at Precious Metal Utilization
   and Recovery
      A. W. Castillero   77
Secondary Refining of Gold and Silver — The New Johnson Matthey
   Chemicals Facility
      F. T. Embleton   81
Gold Recovery by Reduction of Solvent-Extracted Au (III) Chloride Complex
   — A Kinetic Study
      G. F. Reynolds, Stephen G. Baranyai and Leonard O. Moore   93
Oxygen Refining of Smelted Silver Residues
   T. S. Sanmiya and R. R. Matthews   101
Some Aspects of Precious Metals Recovery in the Jewellery Manufacturing
   Industry
      Fernand Verschaeve   113
A Survey of Precious Metal Conservation Techniques in Electroplating
   Operations
      Joseph J. Werbicki, Jr.   117

Analysis

Inductively Coupled Plasma (ICP) in a Precious Metal Laboratory
  B. Horner   125
Developments in Iridium Analyses
  A. Manson and St. J. H. Blakeley   131
Pressure Dissolution — X-Ray Fluorescence Determination of Platinum Metals
  R. E. Price   137

Economics

Excitement in the Metals Markets — A Banker's View
  P. C. Cavelti   147
Platinum Group Metals — Commodity Studies
  T. P. Mohide   157

Minting

The Gold Maple Leaf
  Denis M. Cudahy   165
Proof Coining of Silver
  Joseph Kozol   175
Proof Coining of Karat Gold and Platinum
  J. W. Simpson   183

New Developments

New Developments on the Use of Gold in Electronics and Communications
  Robert R. Davies   193
New Developments in Uses of Silver
  Richard L. Davies   201
New Developments in the Use of Gold in Coins, Medals and Medallions
  David U. Groves   209

Jewellery

Platinum in Jewellery
  D. E. Lundy   219
A Gold Jeweller's Nightmare: IMF — SDR — FED — London Fix —
  Krugerrands — LIFO — FIFO — "EMU" — "SWAPS" — M_1, M_2, M_3, Etc. —
  FIAT
    Paul W. Nordt, Jr.   223
Problems in Gold Ring Casting
  Henry Peterson and Lawrence Diamond   227
Static Vacuum Assist in Casting of Silver and Gold Alloys
  Albert M. Schaler   235
Phase Relations in Cu-Ag-Au Ternary Alloys
  H. Yamauchi, H. A. Yoshimatsu, A. R. Forouhi and D. de Fontaine   241

Dental

The Corrosive Attack of Gold-Based Dental Alloys
  R. M. German, D. C. Wright and R. F. Gallant   253
Physical and Clinical Characterization of Low-Gold Dental Alloys
  K. F. Leinfelder, R. P. Kusy and W. G. Price   259
Consideration of Some Factors Influencing Compatibility of Dental Porcelains
  and Alloys, Part I: Thermo-Physical Properties
    R. P. Whitlock, J. A. Tesk, G. E. O. Widera, A. Holmes and E. E. Parry   273

Consideration of Some Factors Influencing Compatibility of Dental Porcelains
and Alloys, Part II: Porcelain Alloy Stress
   J. A. Tesk, R. P. Whitlock, G. E. O. Widera, A. Holmes and E. E. Parry   283
Corrosion and Tarnish of Ag-In and Ag-In-Pd Alloys
   T. K. Vaidyanathan and A. Prasad   293

Security

Internal Control and Security
   Frederick F. Schauder   303
Meeting the Challenge of the Rising Crimes Against Precious Metal Facilities
   T. L. Weber   305

International Precious Metals Institute

Foreword

THE HISTORY OF THE INTERNATIONAL PRECIOUS METALS INSTITUTE

The International Precious Metals Institute began as an outgrowth of a 1970 graduate offering in the Metallurgy Department of New York University. The course lecturers, Edmund M. Wise, Raymond F. Vines, and John P. Nielsen, emerged as founders of the new entity. Following organization of a steering committee, the first meeting took place on March 5, 1974.

Subsequent meetings ensued producing a mailing list used to solicit 125 charter members. This select group held its first meeting at the Princeton Club on November 18, 1976. Formal business conducted at that time included election of officers, adoption of a constitution, and inclusion of one technical program, a speech on the Technology of Precious Metals Over the Last 4,000 Years, delivered by Dr. Christof Raub of Schwabisch Gmud, West Germany. Other milestones of note included the publication of the first newsletter in June 1977 and compilation of the first membership directory.

The decision to become an international organization evolved from previous independent efforts by members Quentin Dietz and Myron E. Browning. Since substantial evidence indicated foreign interest, the first international conference of IPMI was held May 18, 1977 at the World Trade Center, New York City. Five technical papers were presented at this meeting and officers from the previous year were reelected. The first seminar of the institute, "Sampling and Assaying of Precious Metals" was held at Morristown, New Jersey in April, 1977. This seminar drew a capacity registration and significantly produced a source of income for the growth of the institute. The published proceedings, in soft cover, was the first book of the institute. The entire stock sold out within a few weeks.

Approximately one year later in May, 1978 the second international annual meeting was held. This, the first of the three-day annual conferences, included a program of 35 technical presentations. At this meeting, Dr. Ernst Raub of West Germany was honored for his 50 years of precious metals research and was presented the first International Precious Metals Institute Achievement Award.

The second seminar, entitled "Economic Aspects of Precious Metals" was held in November, 1979 in New York. It included a tour of the Commodity Exchange Center and the headquarters of the Mocatta Corporation. The seminar proceedings were published as our second soft cover book.

College credit course, Noble Metals Metallurgy, was offered in the Fall of 1979 at the Polytechnic Institute of New York. Twenty-five students, half of them auditors or non-credit registrants, participated. Our annual directory indicated that IPMI ended its second year with a 40% increase in membership over the previous year: there were 11 Patrons, 60 Sustaining Members, and approximately 300 individual members.

Our Third Annual International Conference was held in Chicago, May 8 through 10, 1979. Our first venture outside of New York included two social receptions; one sponsored by a group of companies, as is the custom at our annual conferences, and a special one sponsored by Simmons Refining Company. John S. Forbes, Deputy Warden and Assay Master of the Worshipful Company of Goldsmiths of London, received the Achievement award for this year.

Our seminars are proving successful and the two thus far given in New York and Morristown were repeated. They included "Economic Aspects of Precious Metals" in Chicago in November, 1979, and "Sampling and Assaying" in San Francisco in March, 1980. Both seminar proceedings were published.

The Fourth Annual International Conference was held in Toronto, Canada June 2 through 5, 1980. For the first time, a ladies program was offered which proved very successful. The Achievement Award for 1980 was presented to Dr. Leslie B. Hunt, Johnson Matthey & Co. Ltd., London, now retired. The technical program included papers on mining, jewelry, new developments, analytical, minting, economics, recovery and refining, casting, dental, and security. The proceedings of the conference are contained in this publication.

1980 saw a major historic volatility of the prices of precious metals that were experienced throughout the world. Gold peaked at $875 per troy ounce, platinum at over $1,000, and silver at $50. Not only were these historic highs, but they were unprecedented fluctuations - as much as $100 per day per ounce. The price of gold on the free market has been recorded from the year 1200 (in shillings per ounce). At no time in the last 800 years have there been swings in the daily and weekly price per ounce as occurred in the early months of 1980. It appears that IPMI's fortuitous launching afforded an opportunity for the members of the precious metal community to communicate on important matters of mutual interest.

Forty Years of Platinum: Some Recollections and Reflections

Dr. Leslie B. Hunt

Johnson Matthey & Co. Ltd., London, England (Retired)

The President, Ladies and Gentlemen:

First of all, I have to express my great appreciation of the honour you have done me in bestowing upon me your Distinguished Achievement Award for 1980.

Secondly, I have to apologise to you all for my not being able to be present to receive this award owing to certain infirmities that inevitably come with advancing years.

In so far as I may have contributed something of value to the precious metals industry to deserve this award, I feel I must ask you to allocate a share of the merit to the great company with which I have been associated for more than forty years. This remarkable organisation, in which it is a pride and a pleasure to work has, of course, provided me with the opportunities and the stimulus to play some part in the rapidly growing technology of platinum and its allied metals.

When I first joined Johnson Matthey in late 1937, it was already 120 years old and had been engaged in platinum refining and fabrication, originally of course on a very small scale since 1817.

It seems to me therefore appropriate that I should say a few words about the major, and sometimes unexpected, developments that have taken place in the platinum industry over these past forty years, or rather more. These rather personal recollections and reflections will concern the considerable expansion in those applictions of platinum that were already established, but will deal rather more with applications that were unimaginable at the beginning of the period, although the seeds of some of them had been sown very many years earlier. There will therefore be one or two perhaps rather surprising "flashbacks" to the inventive genius of scientists of much earlier times whose ideas took a century or more to find commercial applications.

Let us first look at the simple economic factors. In 1938 the price of platinum was $35 an ounce. In that year the output of platinum from our refineries stemming from Rustenburg Platinum Mines, which began operation in 1931 only to close down for a year in 1932 because of lack of demand, was a mere 25,000 ounces. The present price of platinum is well known to you all; the throughput from the same source has risen year by year until it amounts nowadays to over one and a quarter million ounces and is likely to continue growing.

For a metal that resists virtually all forms of corrosive attack, and one that is consistently recycled from user to refiner and back again, this fifty-fold increase in demand clearly represents a series of quite astonishing developments.

1

A few old established uses have certainly given place to newer forms of corrosion resistant materials, but a considerable number of new applications have obviously made their appearance in this period to transform platinum from its place as a laboratory material, for use as electrical contacts and for a handful of catalytic purposes into the commanding position it now holds as a crucial element in our lives - in energy conservation, in combating pollution, in our communication systems and even in securing our supplies of food.

One of the first developments with which I was involved was concerned with the platinum: rhodium-platinum thermocouple. In 1937 its use was largely confined to heat treatment processes and laboratory work, but in that year a reliable method of determining and controlling the temperature of a huge bath of molten steel in an open-hearth furnace was devised at the British National Physical Laboratory. This "quick immersion" method served to measure temperatures of around 1600 degrees C in a furnace through which flames and gases were roaring at around 1800 degrees C and at a speed of about sixty miles an hour. The technique was adopted with alacrity by the steel industry and has ever since had a profound effect upon the quality of their products.

Another event in 1937 was the development in our research laboratories of a new electrolyte for bright platinum plating to replace the older solutions that had given only a matt deposit. Although this was successful enough, it was soon overtaken by the advent of rhodium plating to give deposits of superior brightness and hardness. Platinum plating has, however, returned in a new form in recent years by replacing the aqueous electolytes with a bath of molten cyanides. By this means, much thicker deposits of platinum having greater ductility and freedom from porosity are now being applied more particularly to some of the refractory metals such as molybdenum and tungsten for use at elevated temperatures.

Another high temperature application of platinum that had just begun to show signs of substantial growth was in the glass manufacturing industry. A hundred and fifty years ago, Michael Faraday had shown that "this beautiful, magnificent and valuable metal", as he described it, provided the ideal container material for the melting of optical glass, avoiding the pick-up from refractories that had hitherto contaminated the product. Many years passed before this suggestion was taken up by the glass industry, but eventually their fire-clay crucibles were replaced by platinum and today many types of glass products including spectacle, camera lenses and television tubes are made in platinum equipment, while continuous processes more recently developed employ tank furnaces completely lined with platinum.

The production of glass fibre involves one of the most exciting uses of platinum. Molten glass at a temperature of around 1300 degrees C flows rapidly through a series of small orifices in a rhodium-platinum alloy bushing, these orifices having to retain their size and alignment over a long period. No other material could retain its mechanical strength and durability in these conditions.

A recent extension in this field that could not possibly have been foreseen forty years ago is the concept of "fibre-optics". It was in 1966 that I.T.T. engineers in England perfected a technique of using glass fibres to transmit light pulses that could be decoded at the receiving end, while in 1971 Corning Glass in America first produced a high purity glass fibre capable of being used more effectively than copper wires in telecommunication systems and of carrying much more information. Quantity production of optical glass fibre, in which the British Post Office has now taken a leading position, is dependent upon a special high-purity grade of platinum for the melting and drawing equipment.

Again in the field of telecommunications and electronics, with the advent of the laser the need arose for large oxide single crystals to be prepared from a melt and it was quickly established that platinum was the most suitable material of construction for the apparatus to be used, although more recently iridium equipment has also been found successful.

The growth of the microelectronics industry has also been largely responsible for the development of a range of metallising preparations based upon platinum, palladium and ruthenium to provide thick films on ceramic substrates.   These consist of the powdered precious metal or metals mixed with finely ground particles of glass and suspended in an organic medium for appliction by dipping or by screen printing on selected areas and then firing. They owe their original concept to the famous German chemist, Martin Klaproth, who, working with the Berlin Porcelain works in 1788, first succeeded in decorating porcelain with such a preparation. This was in fact one of the very earliest applications of platinum in manufacturing industry. It was quickly adopted by the chemists at Sèvres in Paris and then by the early Staffordshire potteries in England, including Spode and Wedgewood, for their so-called silver lustre ware.

Naturally, rather more scientific methods are employed nowadays in the compounding of these preparations.

Another modern application of platinum that owes its genesis to a famous scientist, this time of the early nineteenth century, is the technique of cathodic protection of steel structures and vessels. In 1824, Humphry Davy was consulted by the British Admiralty who were concerned at the "rapid decay of the copper sheeting of His Majesty's ships of war". He proposed the attachment of a small piece of zinc to nullify electrochemical action on the copper sheathing, a treatment that proved successful when adopted on a wooden warship of the time. Davy also investigated the impressed current system, but did not then consider it to be practical in service conditions, no doubt because reliable batteries had not yet been developed.

It was not, however, until 1956 that the United States Navy seriously began to experiment with platinum clad titanium anodes for the protection of their ships and submarines. A great deal of development work had to be undertaken by those concerned in the design and construction of cathodic protection systems, and in more recent years niobium has replaced titanium as the substrate metal, but today this invaluable means of avoiding corrosion is an established technique on many types of ships and steel structures, including off-shore oil rigs - themselves undreamt of forty years ago!

Much has been written about the fuel cell as a source of power combining high thermal efficiency with very low risk of environmental pollution.   After many difficulties and extensive research programmes, commercial megawatt generating systems are now in the course of installation in the United States, these relying upon a coating of platinum dispersed on carbon applied to the electrodes to activate the electrochemical reaction that converts hydrogen or a hydrocarbon and oxygen into electrical energy. Origin of the fuel cell goes back, however, as far as 1842, when the London scientist, W.R. Grove, published a paper on "A Gaseous Voltaic Battery" in which he described the first practical fuel cell constructed from platinum foil coated with spongy platinum and with dilute sulphuric acid as the electrolyte.

Fuel cells containing platinum were employed as sources of power in the Apollo and Gemini spacecraft - once more a development inconceivable forty years ago - but for longer missions, such as the Viking voyages to Jupiter and Mars and for satellites, another source has been employed. This consists of a radioisotopic thermoelectric generator in which plutonium 238 provides heat; this heat being converted to electricity by a number of thermocouples. It is of course necessary to encapsulate the plutonium - which may reach a temperature approaching 2000 degrees C - in absolute safety under both normal and accidental conditions, and for this purpose special platinum alloys containing both rhodium and tungsten have so far been found most suitable. In this development, Dr. Chain T. Liu, played a significant part in the research at the Oak Ridge National Laboratory.

But by far the greatest transformation over this period of years has been in the field of catalysis. Forty years ago, apart from the established use of rhodium-platinum alloys in the form of finely woven gauzes for the catalytic oxidation of ammonia to nitric acid (and thence to the production of vital fertilisers and the enhancement of our food supplies) catalyst production was modest and comprised limited amounts of palladium on charcoal, mostly for the pharmaceutical industry, and declining quantities of platinum catalysts for the production of sulphuric acid by the contact process, these already being replaced by vanadium pentoxide.

The major breakthrough into the petroleum and petrochemical industries occurred in 1949 when Universal Oil Products installed their first catalytic reformer - or Platformer as they named it - for the upgrading of low octane naphthas to high quality products and for the extraction of benzene, toluene and xylenes. The first installation required 400 ounces of platinum at $70 an ounce, an amount considered quite astonishing by many observers, even in the oil industry. Today, that industry takes over 100,000 ounces a year to service the several types of reforming processes that have since been developed.

Probably the best known appliction of a platinum catalyst to the general public outside the precious metals industry is also directly associated with petroleum products. I refer, of course, to the extensive use of platinum now being made in the control of automobile exhaust gases. This is too well known to all of you to require description, but an interesting point arises here in regard to supply and demand. Legislation in the United States called for a considerable reduction in noxious emissions from automobiles produced in the year 1975, with even higher standards to be expected in subsequent years. Once research had established that platinum metal catalysts were capable of meeting these requirements the implications were clear that a considerable increase in their production would rapidly be needed. Now any major increase in the mining, extraction and refining of the platinum metals necessarily involves long forward planning and substantial capital expenditure.

The mineral extracted at Rustenburg contains only about one-eighth of an ounce of platinum per ton. While the reserves are enormous, obviously eight tons of rock have to be broken and brought to the surface to yield just one ounce of refined platinum. Improvements in mining technology and the development of more rapid extraction and refining methods have certainly reduced the lead time in recent years, but it remains true to say that rapid expansion of production to meet a new, short term or unforeseen demand is not readily acceptable. Fortunately, the automobile companies were most co-operative in discussing their likely demands well ahead of time, with the result that the needed quantities of platinum and of one or two of its associated metals were able to be made available by the producers. This co-operative attitude is becoming more and more advisable in order to ensure that platinum producers and refiners can respond willingly and effectively to the changing and growing needs of modern technology throughout the world.

The other major application of platinum metal catalysis lies in the field of pollution control or abatement. From the manufacture of nitric acid and the elimination of noxious tail gases with a platinum catalyst to many of the more everyday industrial processes such as plastics manufacture, printing, wire enamelling, paint drying and many others, catalytic combustion systems can now eliminate many of the toxic gaseous waste products that were formerly a cause of unpleasantness, irritation, or even disease.

Returning for a moment to the automobile engine and other forms of motive power, it has recently been demonstrated that a metal-supported platinum catalyst can be used to fire gas turbine engines, with the advantage that they then emit less pollutants than conventionally flame-fired engines.

One final reflection that harks back almost, but not quite, forty years. On May 15, 1941 on a Royal Air Force Station in England, the world's first jet propelled aircraft designed by Sir Frank Whittle, took off and flew for seventeen minutes, thus inaugurating a completely new phase in air travel. This first engine was fitted with stainless steel turbine blades but these were quickly found to be inadequate and led to the development of the so-called Superalloys, based upon the nickel-chromium system, by the International Nickel Company. These have been consistently improved over the years to permit higher and higher operating temperatures and so increasing thrust until a maximum seems to have been reached.

In the last few years, however, the development of a platinum coating process on the blades has increased their durability, while the most recent research carried out in the Johnson Matthey Research Centre has shown that small additions of one or other of the platinum metals to some new nickel-base Superalloys have greatly improved their resistance to oxidation and corrosion in severe environments at high temperatures, while retaining their good mechanical properties. Although these alloys will probably not be required in aircraft engines, operating in clear atmospheres, they offer a marked advance for land-based turbines for power generation and in other industrial uses, and their development seems to me to be indicative of the importance of platinum's remarkable resistance to attack at high temperatures in modern technology.

Mining and Extraction

Placer Mining in Alaska

Earl H. Beistline

School of Mineral Industry
University of Alaska
Fairbanks, Alaska 99701

ABSTRACT

A review of the location of gold placer districts in Alaska, unique conditions encountered in many northern deposits, description of typical mining methods used, current interest and challenges for the future of the gold placer mining industry. The information is presented in an integrated manner to give a description of major facets of Alaska's placer mining industry: Alaska, gold, mining districts, recovery, regulation.

KEYWORDS

Alaska; gold; mining districts; mining methods; recovery; regulation.

INTRODUCTION

Gold has entranced mankind since the days of the Egyptian Pharaohs. Because of its enduring beauty, relative scarcity, and resulting high value, people have hoarded it over the centuries as a means of security. Nations have fought wars to secure gold and to reap the benefits it brought to their countries.

Within the last 100 years, men and women were drawn to Canada and Alaska in search of precious metals. Many of these pioneered in exploring the north country and laid the foundation for some of the cities and societies we know today.

Generally, gold is found in two types of deposits--lode and placer. Lode deposits are those found in place in solid rock. Lode mining is done by breaking the material and processing it through a mill to recover the precious metals. On the other hand, placer deposits are often formed by the erosion of lode deposits and the resulting concentration of the gold in the gravels of stream beds.

A typical vertical section of a placer deposit in a stream valley would consist of a layer of surface vegetation underlain by windblown loess (muck overburden) which in turn is underlain by gold bearing gravels resting on bedrock.

Gold is not found in the overburden but in the stream gravel, sometimes dispersed through the gravel but in some deposits concentrated on and within a few feet of

bedrock. Gravels may have a maximum depth of 100-150 feet, but many placer
deposits have depths of 10-50 feet.

The early-day stampeders searched for gold placer deposits which, once found,
required a minimum of equipment to mine and in which the valuable product could be
recovered in a simple washing plant.

The Klondike discovery near Dawson in the Yukon Territory in 1896 by George Carmack
is well known. It has been dramatized by many authors including Pierre Berton in
his excellent account (The Klondike Fever). This discovery did more than any other
single event to stimulate people to penetrate into the Yukon Territory and Alaska
to search for gold. Before and following the Klondike discovery, numerous other
placer areas were found, including Nome in 1898 and Fairbanks in 1902. As a result,
the gold mining industry of Alaska was well on its way.

LOCATION OF GOLD PLACER DISTRICTS - ALASKA

Alaska is abundantly endowed with placer gold deposits. These extend from the
eastern boundary near the gold deposits of the Yukon Territory through the width of
Alaska to its western boundary and beyond into Siberia.

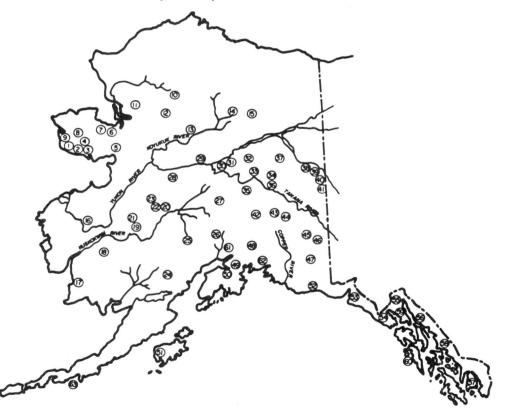

Fig. 1.   Alaskan gold camps and gold production, showing the wide distribution of
           gold deposits within the State. (from Robinson, M.S. and Bundtzen, T.K.,
           1979, Historic gold production in Alaska - A "minisummary": Alaska
           Division of Geological and Geophysical Surveys, Mines and Geology Bulle-
           tin, Vol. XXVIII, No. 3, September, 1979.)

Figure 1 shows the location and distribution of Alaska gold camps. Table 1 gives the names of the camps, approximate gold production and date of discovery of specific camps. The majority of production has been from placer operations, although some is from gold lode deposits such as #56, Juneau: #60, Chichagof; and #61, Willow Creek. A study of Fig. 1 reveals major placer concentrations in the Seward Peninsula (#1-9), the Kuskokwim River area including the Iditarod and Innoko (20-23), the Interior of Alaska and the Fortymile (38-41). The Fairbanks area has had the greatest production of all the camps in Alaska. Goodnews Bay (#17) was a major platinum producer for approximately 45 years. The company terminated its platinum placer dredging operation in 1974.

TABLE 1. Tabulation of Gold Production from the Gold Camps in Fig. 1, and Date of Discovery of Each Camp.
(from Robinson, M.S. and Bundtzen, T.K., 1979, Historic gold production in Alaska - A " minisummary ": Alaska Division of Geological and Geophysical Surveys, Mines and Geology Bulletin, Vol. XXVIII No. 3, September, 1979.)

No.	Camp*	Gold Production (ounces)	Discovery Date	No.	Camp*	Gold Production (ounces)	Discovery Date
1.	Nome	3,606,000	1898	33.	Fairbanks	7,464,200	1902
2.	Solomon	251,000	1899	34.	Chena (included in Fairbanks Production)		
3.	Bluff	90,200	1899	35.	Bonnifield	45,000	1903
4.	Council	588,000	1898	36.	Richardson	95,000	1905
5.	Koyuk	52,000	1915	37.	Circle	730,000	1893
6.	Fairhaven (Candle)	179,000	1901	38.	Woodchopper - Coal Creek(included in Circle production)		
7.	Fairhaven (Inmachuck)	277,000	1900	39.	Seventymile (included in Fortymile production)		
8.	Kougarok	150,400	1900	40.	Eagle	40,200	1895
9.	Port Clarence	28,000	1898	41.	Fortymile	400,000	1886
10.	Noatak	9,000	1898	42.	Valdez Creek	37,000	1903
11.	Kobuk (Squirrel River)	7,000	1909	43.	Delta	2,500	
12.	Kobuk (Shungnak)	15,000	1898	44.	Chistochina-Chisna	141,000	1898
13.	Koyukuk (Hughes)	201,000	1910	45.	Nabesna	63,300	1899
14.	Noyukuk (Nolan)	290,000	1893	46.	Chisana	44,800	1910
15.	Chandalar	30,700	1905	47.	Nizana	143,500	1901
16.	Marshall (Anvik)	120,000	1913	48.	Nelchina	2,900	1912
17.	Goodnews Bay	29,700	1900	49.	Girdwood	125,000	1895
18.	Kuskokwim (Aniak)	230,600	1901	50.	Hope (included in Girdwood production)		
19.	Kuskokwim (Georgetown)	14,500	1909	51.	Kodiak	4,800	1895
20.	Kuskokwim (Mckinley)	173,500	1910	52.	Yakataga	15,700	1898
21.	Iditarod	1,320,000	1908	53.	Yakutat	2,500	1880
22.	Innoko	350,000	1906	54.	Lituya Bay	1,200	1894
23.	Tolstoi	87,200		55.	Porcupine	61,000	1898
24.	Iliamna (Lake Clark)	1,500	1902	56.	Juneau (Gold Belt)	7,107,000	1880
25.	Skwentna (included in Yentna production)			57.	Ketchikan-Hyder	62,000	1898
26.	Yentna (Cache Creek)	115,200	1905	58.	Sumdum	15,000	1869
27.	Kantishna	55,000	1903	59.	Glacier Bay	11,000	
28.	Ruby	389,100	1907	60.	Chichagof	770,000	1871
29.	Gold Hill	1,200	1907	61.	Willow Creek	652,000	1897
30.	Hot Springs	447,900	1898	62.	Prince William Sound	137,900	1894
31.	Rampart	86,800	1882	63.	Unga Island	108,900	1891
32.	Tolovana	375,000	1914				

(Data compiled from U.S. Geological Survey publications, U.S. Bureau of Mines records, Alaska Division of Geological and Geophysical Surveys records and publications, Mineral Industry Research Laboratory research projects and other sources.)

*Camp names are those that appear in official recording district records. Many are also known by other names, some of which are shown in brackets.

More detailed information and references are found in Map MR-38, Placer Gold Occurrences in Alaska, compiled by E.H. Cobb, U.S. Geological Survey, Department of the Interior, Washington, D.C.

Many of the locations shown had major gold production near the turn of the century. Production declined during and after the first world war, but increased with the advent of large scale dredging in the late 1920's. For a number of years after World War II it was at a low ebb because of the frozen price of gold and the increased cost of mining. However, during the past five years, a great interest has been shown in gold mining because of the high price and, consequently, many of the placer districts are being revitalized as new mining operations begin.

UNIQUE NORTHERN CONDITIONS

Early-day placer miners and prospectors coming to Alaska and the Yukon Territory found that placer deposits were found in new and unusual physical conditions. First, many of the deposits were covered with overburden reaching depths of more than 100 feet. This material, for the most part, was wind-blown loess containing some partially decayed vegetation and the remains of Pleistocene animals. No gold was in the overburden which is often referred to in literature as "muck".

The second unusual condition was that many of the placer deposits were permanently frozen from the surface to bedrock and below. The frozen muck contained large quantities of ice in the forms of lenses, wedges and small particles dispersed throughout the muck. The underlying frozen gold bearing gravels require thawing before mining could be accomplished.

The challenges of finding and mining gold under these conditions were accepted and solved by the early-day miners. Their efforts and solutions have led to the development of modern mining methods to meet the characteristics of the frozen deposits.

TYPICAL MINING METHODS

Placer mining methods in Alaska and the Yukon Territory evolved by men applying their mining knowledge to meet the Northern conditions and the necessity of obtaining the precious metals in an economically sound mining operation.

Initially, shafts were sunk to the paystreak by thawing the frozen ground with wood fires, hot rocks and finally steam. The steam technique was further advanced into an underground placer mining method known as drift mining. Open pit methods developed using various types of scrapers and dredges.

In the early 1930's, placer mining techniques were greatly advanced by the introduction of the lightweight diesel engine. This allowed the development of the bulldozer, the dragline, and more efficient pumping units, all of which led to more efficient mining methods. In recent years, the bucket loader has found an increasing use in placer operations and much experimenting has and is being done to increase the efficiency of recovery plants and to retain more of the fine gold that has been lost in some operations in the past.

SMALL-SCALE RECOVERY METHODS

Week-end miners and hobbyists have joined the search for gold. The familiar gold pan is used most often, but small sluice boxes of various designs, usually fed by hand shoveling, are also used. In addition, rockers and various small-scale mechanical recovery units are used in an attempt to "mine for gold." In recent years, suction dredges have become popular, with many individuals carrying on quite successful small-scale prospecting and mining.

HYDRAULIC MINING

Hydraulic mining is defined as using water under pressure through nozzles to remove overburden, move the gold-bearing gravel to the sluice box and to remove the tailings discharged from the sluice box. True hydraulic mining requires a large amount of water that preferably can be brought in by gravity ditch with heads of up to several hundred feet. In Alaska only a few hydraulic mines are in operation generally with very reasonable mining costs. Sluice boxes may have widths up to five

feet, lengths of 50-70 feet, and grades up to 1-1/2 inches per foot of length.
Recently steeper grades, up to 3 inches per foot, have been used. Riffles are most
often of the hungarian type and may be made of angle iron or railroad rails.

MECHANICAL METHODS

Mechanical mining methods are the most common of the type used. Such methods use
combinations of various machines to meet the characteristics of individual deposits.
These machines include bulldozers, draglines, loaders, conveyor belts, pumps and
various recovery plants.

A typical operation involves locating the paystreak by drilling, trenching and
sinking shafts or pits. Muck overburden has been removed in the past by water, but
mechanical means are being increasingly used. Barren gravel is stripped with
bulldozers and in some cases with mechanical scrapers or draglines.

A source of water is necessary and this may be obtained by ditches and pipelines
that bring water to the recovery plant with a gravity head or by pumping from a
nearby source. In some cases where makeup water is in short supply, a dam may be
constructed downstream and the water returned to the washing plant by pumping.
Gravel is fed to the recovery plant at a rate that gives a uniform flow of water
and material through the unit without surging or plugging. A drain must be dug to
allow the cut to be kept dry and provide drainage at the sluice box. In addition,
a settling pond may be necessary to meet environmental requirements or to save
water for recirculating. Usually such a pond will require cleaning periodically.

Gold is removed from the sluice box through a "cleanup" procedure which separates
the gold from the sand through a paddling process in the sluice or by removing all
gold and heavy black sands from the sluice and separating the constituents in an
auxiliary plant. Figure 2 shows a typical layout in which bulldozers, dragline,
sluice box, water from a ditch and makeup water from the creek are used.

DREDGING

In the past, large bucket line dredges were used in a number of areas in the Yukon
Territory and Alaska, but at present, only a few dredges are operating during the
mining season in Alaska. Dredge operations in the past involved prospecting the
ground to establish limits of pay, obtaining sufficient water for mining by construction of ditch systems and/or pumping plants and obtaining a source of power.
The actual operation began with the removal of muck by hydraulic methods, sometimes
to depths of 100 feet. The gold bearing gravels were then thawed to bedrock by
using the Miles cold water thawing system of introducing cold water through pipes,
and finally the excavation of the gravel by a floating bucket line dredge which
contained equipment for the separation and recovery of the precious metals right on
the dredge.

A typical dredging section with elements of stripping, thawing and dredging are
shown in Fig. 3.

DRIFT MINING

This involved sinking a vertical shaft to bedrock through the frozen muck and
gravel which, in some cases, was as deep as 200 feet. In excavating the shaft the
ground was thawed by wood fires and hot rocks in the more primitive operations, and
steam was used in later larger scale mines. The thawed material from the shaft was

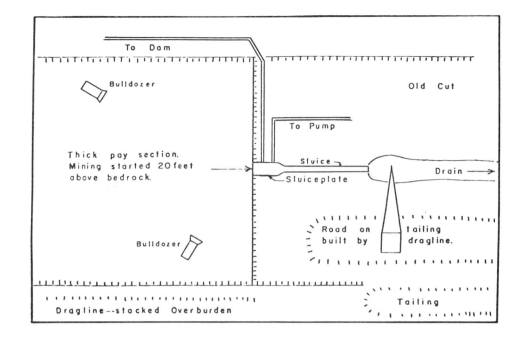

Fig. 2.   Plan view of mine.   (from U.S. Bureau of Mines Information Circular
           7926.)

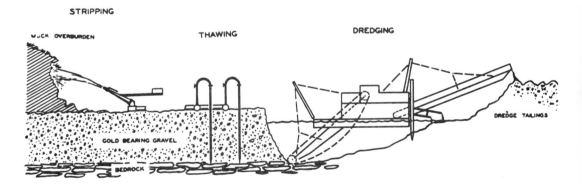

Fig. 3.   Gold dredging in Fairbanks district, Alaska.   Compliments of the United
           States Smelting Refining and Mining Company.

hoisted to the surface by a steam hoist and discarded.   The shaft, often 8' x 8' in
plan view, was timbered for protection of the shaft and the personnel in it.
Drifts were driven both upstream and downstream to lengths of possibly 250 feet in
each direction.   From the ends of the drifts, crosscut openings were driven at
right angles out to the limits of the paystreak which might be 100-150 feet in each
direction.   Mining then retreated toward the shaft.   In the mining cycle, steam

points were driven into the retreating face on one side of the main drift. If points were in short supply they were removed after being driven home, and small pipes, called sweaters, were inserted. Up to six feet was thus thawed during a day of steaming. In the interim, the thawed gravel from the previous day's thawing was picked loose, loaded into wheelbarrows, transported to the shaft and hoisted to the surface. The pay gravel was washed through a sluice box in summer or stockpiled in winter.

At present, increased interest is being shown in drift mining although vastly improved techniques would be used.

CURRENT INTEREST

Current interest in placer mining is at a high level because of the increased price of gold. This was well shown by the large number of people attending two placer mining conferences in April 1980.

The Whitehorse conference, put on by the Department of Indian and Northern Affairs, Yukon Territory, Canada, drew more than 350 and was a sellout. The School of Mineral Industry-Alaska Miners' Association Placer Mining Conference at the University of Alaska, Fairbanks, had an attendance of approximately 500 people.

The majority of people now interested are of the family or partnership type (2-5 persons); however, there will be some larger scale mining by large companies this year.

Methods used will, for the most part, be of the mechanical type and will consist of numerous arrangements and combinations of equipment. Recovery plant design is attracting considerable attention by a number of miners and, as a result, a large variety of sluice boxes and auxiliary accessories will be used this year. The object of updating recovery plants is to wash more yardage on a daily basis and to recover more fine gold.

CHALLENGES FOR THE FUTURE

Numerous challenges exist in further enhancing the efficiency and economy of the placer mining industry in Alaska. Some of these are:

1. Refinement of exploration techniques to allow faster and accurate information to be obtained.

2. Continued development of recovery units, especially directed toward fine gold recovery and larger daily yardage.

3. Development of reasonable techniques to meet various requirements of environmental, ecological, reclamation and "pollution" regulations and laws.

The opening stanza of Robert Service's "Trail of Ninety-Eight" is as pertinent as ever.

> "GOLD! We leaped from our benches. GOLD! We sprang from our stools.
> GOLD! We whirled in the furrow, fired with the faith of fools.
> Fearless, unfound, unfitted, far from the night and the cold,
> Heard we the clarion summons, followed the master lure--GOLD!"

An Overview of Gold and Silver Mining in Canada

M.R. Brown, P. Eng.

The Northern Miner, Toronto, Canada

ABSTRACT

The vigorous mining industry that exists in Canada today started with the chance discovery of rich silver deposits at Cobalt, Ontario at the turn of the century. Although the output of both gold and silver has been dropping in recent years, these metals continue to play a key role, especially gold of which we are the world's third largest producer. Output of that metal peaked in 1941 at 5,345,179 ounces from 144 producers.

Last year there were only 21 operating gold mines that turned out just over 1,600,000 ounces. But with the sharp rise in the price of these metals, offering new incentive, the search for new deposits is rising, with production once again on the increase.

A major new discovery at Detour Lake in Northeastern Ontario is now under full development and is shaping as Canada's largest gold deposit. But, the greatest potential for future expansion of the gold mining industry is thought to exist in the country's northern frontiers, especially the Northwest Territories where several new high grade mines are now being readied for production.

INTRODUCTION

Both gold and silver have long played a key role in Canadian mining and in all likelihood will continue to do so . . . . especially gold.

It was, in fact, the discovery of native silver on August 14, 1903 by workmen pushing the Temiskaming and Northern Ontario Railway northward from North Bay, that started the ball rolling and launched the vigorous mining industry we have in this country today. That find proved fabulously rich and almost overnight resulted in the birth of our famed Cobalt Camp which, in just a few short years, turned out half a billion ounces of the white metal that in turn opened the gate to our Precambrian Shield. It was the profits from that exciting silver development that financed the prospecting sorties that resulted in the Porcupine, Kirkland Lake, Noranda and Flin Flon camps.

While there are still some half dozen "straight" silver mines operating in this country, production of this metal is now largely a by-product of our base metal mines, led by Texasgulf which turns out some 7 to 8 million ounces anually from its Kidd Creek Mine at Timmins (Porcupine). Last year Canada's silver output totalled 38,100,000 ounces, down 4 percent from the previous year.

But it is on gold that I would like to dwell. Canada, as most of you are probably aware, is the world's third largest producer of this metal following (well behind) South Africa and the Soviet Union. Our gold output peaked in 1941 when we boasted no less than 144 producers which turned out 5,345,179 ounces. Understandably, many of these producers were relatively small. But, the wartime drain of manpower and supplies took its toll, resulting in many mine closures. With the post war escallation in costs and the then existing fixed price of gold, production never again approached those peak levels. But, thanks to the Federal Government's Emergency Gold Mining Assistance Act (EGMA) which was put into effect, a nucleus of gold mining operations was kept in business. The value of our gold production continued to lead to all other metals until 1953 when it was finally surpassed by both copper and nickel.

Last year there were but 21 operating gold mines in this country with only 19 mills operating. Production was some 1,600,000 ounces.

But it will be quite a different ball game this time. One of the difficulties, of course, is that prospecting for gold, which is a unique skill and which had been highly developed in this country, was largely lost over the years of retrenchment. Also, our highly developed geophysical and exploration techniques have much less application in the search for gold than for base metals. (An exception is a major new gold discovery at Detour Lake in Northeastern Ontario which I will come back to in some detail later.)

At today's gold price, which is much more attractive in this country than, say, the U.S., because of the lower value of the Canadian dollar, we can now profitably mine down to a grade of 0.04 ounce per ton underground, somethjing undreamed of during our past gold mining boom. An open pit deposit, of course, could be mined lower still. Too, there are much more efficient and labor saving methods today. I am thinking primarily of ramp mining with rubber tired load-haul units versus the higher cost shaft approach.

Also, we are now able to haul ore in large on-highway trucks much longer distances to centralcustom milling plants. A classic example is that of Pamour Porcupine Mines at Timmins which is going all-out in the custom milling business, having raised its daily mill capacity to some 7,000 tons and now drawing low grade mill feed (i.e. 0.1 ounces and less) from mines within a radius of 70 miles of its plant.

So far, the resurgence of gold mining in Canada has been largely restricted to the expansion of existing mines and the reopening of long-closed ones within trucking distance of existing mills. There has as yet been very little new mill construction or development of new mines.

One of the most notable expansions is that at Dome Mines. This veteran Porcupine area mine, which has been in continuous production since 1910 (70 years), is truly one of this country's great gold mines. Last year it milled 663,149 tons. Operating costs were $29.25 per ton or $205.04 per ounce of gold produced. This mine is now in the throes of a $50 million program that will increase milling rate by 50 percent to 3,000 tons daily. It calls for a new 5,400 foot shaft, new grinding facilities and general plant modernization.

But perhaps the most remarkable gold mine in this country is Campbell Red Lake at Red Lake, Ontario. A 56.8 percent owned subsidiary of Dome Mines, it came into being much later, but is already this country's leading gold producer. It is also the highest grade and lowest cost gold mine in Canada. Last year it milled 300,000 tons of ore averaging 0.656 ounces for a recovery of 185,005 ounces valued at $70,055,000. Operating costs were $44 per ton or $71.41 per ounce. This operation is likewise being expanded. Mill capacity will be increased by thirty percent to 390,000 tons per year, with gold production expected to rise fifteen percent to 212,000 ounces by year end. This expansion, it should be noted, was planned prior to the rise in the price of gold. So further expansions at this unique mine are quite possible. (The Dickenson Mine, adjoining Campbell on the east and which finds its ore in the same goldbearing structure and which has been in production since 1948, is likewise expanding its operation . . . from 500 tons daily to 700 tons "for starters".) Cost will be $9 million.

Characteristic of many of the gold mines in this country are the relatively low "official" ore reserves reported by both Dome and Campbell. As of December 31, 1979, Dome estimated its reserves at 1,896,000 tons averaging 0.216 ounces, virtually unchanged from the previous year while Campbell reported 1,196,500 tons averaging 0.657 ounces, an increase of only 77,000 tons during the year. These figures, I feel, are on the conservative side, for both of these mines have very long life expectancies if metal prices hold to anything like present levels . . . at least 25 years in both cases.

As most operators are fully aware, gold deposition in the Canadian-type deposits is frequently erratic, making it difficult to come up with valid tonnage and grade estimates based on diamond drilling and traditional sampling methods. In fact it would be very costly to outline sufficient ore reserves to justify the high capital cost of bringing a new mine into production today, especially if it requires its own milling plant. If one were to add up the published ore reserves of all the operating gold mines in this country, it might appear shockingly low, especially if compared to our other metal mines, a number of which show reserves in the order of 100 million tons.

But, I might point out that gold mines in this country traditionally operate with about a three year ore reserve, probably a spinoff from the long periods when they operated under EGMA which restrictd them to that level. However, we have both gold and silver mines operating with virtually no ore reserves, yet they have a remarkable record of longevity.

The traditional gold mining areas of Northwestern Quebec are currently witnessing quite a resurgence. But again, much of this pertains to the reopening of former marginal mines that didn't quite make the grade before the price increase.

Kiena Gold Mines, with a partially developed gold property in the Malartic Area is an example. Closed since 1965, reserves at that time were reported at 2,767,000 tons grading 0.23 ounces. Recent work suggests this might be increased significantly but of a somewhat lower grade. The mine is now being readied for production (custom milling) at a rate of 1,000 tons daily and at a cost of $22 million. Tentative plans are for this company to erect its own large milling unit several years hence.

But there are some brand new developments in that area as well, with the Little Long Lac organization being particularly active. It has just recently brought two new mines into production (custom shippers) -Thompson Bousquet Gold Mines which is an underground operation with reserves of 775,000 tons grading 0.22 ounces and Silverstack Mines which is an open pit operation with reserves of some 4.4 million tons of open pit and underground reserves grading 0.175 ounces.

Another brand new development is the Ferderber Mine of Belmoral Mines in the Val d'Or area. This discovery was made in what is known as the Bourlamaque batholith in a granodiorite formation that was almost completely shunned by previous gold seekers in that region. This mine just recently commenced shipping ore at the rate of 10,000 tons monthly to a custom mill. Grade is in the 0.20 to 0.25 range. But, the company is now builing its own 1,500 ton milling plant and plans to treat 1,000 tons of ore daily from the Ferderber and the balance from the nearby subsidiary Bras d'Or Mines where a 1,500 foot shaft is now going down.

A relatively new Northwestern Quebec mine and one of that province's leading producers is that of Agnico-Eagle Mines at Joutel. It started in 1974 at 1,000 tons daily. Because of a steadily expanding ore picture at depth, the mill is now being increased to 1,500 tons. This is one of the lower cost operations in the country, last year's average being $129.86 per ounce produced. Grade of this sulphide deposit approximates 0.25 ounces.

But what is considered the most important gold discovery made in this country in years is the one I previously referred to at Detour Lake.

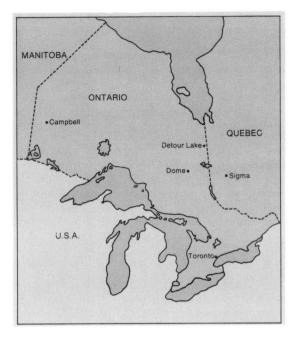

Fig. 1   Detour Lake

Also a sulphide deposit, it was made by Amoco Petroleum Canada Ltd. by geophysical means while flying the area in 1974 in search of base metals following an important discovery made by Selco Explorations on the Quebec side of the boundary in the same greenstone belt.

Extensive subsequent exploratory drilling and limited underground work suggests a deposit containing a minimum of 10 million tons grading approximately 0.20 ounces, making it the largest known gold deposit in this country. It is now under full development by the Dome-Campbell Red Lake Mines' team under a joint venture agreement entered into last year. C.H. Brehaut, Dome's Vice President of Operations, stated recently, "It appears that this property will support a milling plant of at least 2,000 tons per day".

Because of the importance of this new development at Detour Lake, I feel it is in order to quote at some length from the coverage accorded this project in Dome's recent Annual Report.

"Diamond drilling was started by Amoco in October 1974 and, after encouraging values were obtained in the first three holes, an additional 362 claims were staked surrounding the original block. Surface drilling continued through to August 1976 by which time 157,000 feet had been drilled on various geophysical targets. Drilling on the original anomaly accounted for 87 percent of the footage and a major zone was identified which, at 1,800 feet below surface, was still open to depth.

"In October, 1976 a comprehensive study was commissioned which included underground examination of the ore zone, metallurgical test work and engineering studies. The underground work involved 3,000 feet of development and 35,200 feet of diamond drilling. A decline was driven to a level 400 feet below surface, at which point three crosscuts were driven across the ore structure for examination and bulk sampling purposes. Diamond drilling was carried out from a drift driven parallel to the zone in the hanging wall.

"As a result of this program, reserves were estimated at 6.2 million tons averaging approximately 0.19 ounces of gold per ton. It was obvious that a significant discovery had been made but at the price of gold that existed early in 1979 the economics of the project were in doubt. At this point in time, Campbell and Dome were invited to acquire an interest in the property.

"The gold occurrences are found principally in a quartz fracture zone within basaltic flows on the hanging wall side of a chert horizon. Significant gold values have also been found in the chert horizon and in a talc carbonate zone on the footwall side of the chert horizon.

"The outline of the main quartz fracture zone has an indicated strike length of 700 to 900 feet and plunges at 35 degrees from the horizontal. In general, the dip is fairly steep but individual sections, as inferred from diamond drilling, vary from vertical to 45 degrees. The zone has been traced to a vertical depth of 1,800 feet below surface and the results of deep drilling indicate that the main quartz zone is open to depth."

Officials have not intimated what it will cost to bring this sizeable project into production, but in view of its remote location and lack of power, it will obviously be high. I have been at the minesite and venture a "guesstimate" that it will exceed $100 million .... perhaps $150 million.

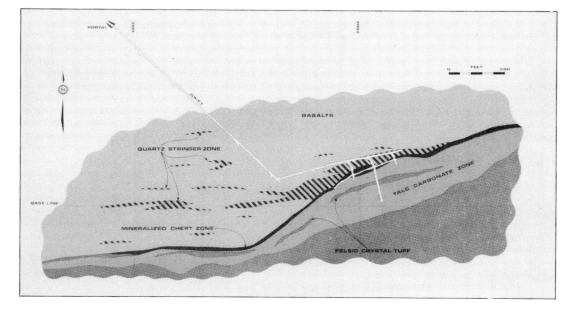

Fig. 2.   Idealized Geological Plan of the Detour Lake Site
at the 400 Foot Elevation

There are several other up-coming potential new producers that are already looming for Ontario. One is Goldlund Mines' property in the Sioux Lookout area now under full development. Idle for years, ore reserves are in the order of 2.0 million tons in the 0.18 to 0.20 ounce range. This would likely support a 1,000 ton mill. Another is the brand new Owl Creek deposit of Texasgulf which lies only two miles from that firm's big Kidd Creek mine at Timmins. Drilling has already outlined some 2.0 million tons averaging 0.15 ounces, with the structure described as "open in several directions".

British Columbia, too, is starting to feel a gold boom coming on. Small, but of excellent grade, that western province's Northair Mines represents its first new gold producer in over 40 years. Also well underway is the construction of a 5,000 ton plant at the Sam Goosly silver-gold property of Equity Silver Mines near Houston, Northwestern B.C. Open pit reserves are estimated at 30,800,000 tons grading 3.10 ounces silver; 0.028 ounces gold and 0.384 percent copper. Cost of this project is estimated at $107 million.

Also scheduled to come into production early next year is the Ladner Creek development of Carolin Mines near Hope, at a rate of 1,500 tons daily. Drill indicated reserves are 1,530,000 tons grading 0.14 ounces gold. Still another interesting new B.C. gold development is that of Consolidated Cinola Mines on the Queen Charlotte Islands on which a production decision is expected before year end. Drilling to date has indicated a minimum of 30 million tons of low grade material running 0.06 ounces of gold with approximately the same amount of silver. A 50 ton pilot mill is now being erected.

But perhaps the greatest potential for future expansion of the gold mining industry in this country lies in our northern frontiers, i.e. the Northwest Territories, although the resurgence there has barely started.

The effect of the higher gold price on operating mines in the north is well demonstrated in the case of Giant Yellowknife Mines. In production since 1949, that veteran N.W.T. operation has milled some 10.5 million tons and extracted 4,940,000 ounces. Until the recent price rise this mine had only about two years of productive life remaining. But because it is now able to mine a much lower grade, large tonnages of sub-marginal material have now been reclassified as ore, putting over five years' reserves in sight. And, it will likely continue beyond that.

At least two small but high grade gold mines are to be brought into production in the N.W.T. . . . . the Camlaren Mine at Gordon Lake and that of O'Brien Resources at Cullaton Lake. There will be others, several of which are just now coming under serious investigation.

In conclusion, you will have sensed that I am a gold optimist. For this I make no apologies. Being a gold miner myself, and having lived with that industry for years, it simply couldn't be otherwise.

I firmly believe that there is a tremendous potential for gold mining in Canada, given a favourable economic and political climate for its development.

Mineralogy and Distribution of the Platinum Group in Mill Samples from the Cu-Ni Deposits of the Sudbury, Ontario, Area

Louis J. Cabri

Mineralogy Section, Mineral Sciences Laboratories, CANMET,
Department of Energy, Mines and Resources, 555 Booth Street,
Ottawa, Ontario, K1A 0G1

ABSTRACT

Detailed mineralogical studies have shown that the platinum-group element (PGE) mineralogy is variable between deposits but, for the Sudbury area as a whole, the principal platinum-group minerals (PGM) are: michenerite ($PdBiTe$), sperrylite ($PtAs_2$) and moncheite ($PtTe_2$). Less common are insizwaite ($PtBi_2$), froodite ($PdBi_2$), sudburyite ($PdSb$), kotulskite ($PdTe$), merenskyite ($PdTe_2$), niggliite ($PtSn$), mertieite II (Pd_8Sb_3), hollingworthite ($RhAsS$), irarsite ($IrAsS$) and three unnamed minerals. *In situ* analyses have also established that Pd, Pt and Rh occur as dilute solid solutions (56-1840 ppm) in sulpharsenides (cobaltite, $CoAsS$, and gersdorffite, $NiAsS$). The presence of PGE as more dilute solid solutions in the common sulphides can only be determined by indirect methods. It is proposed to demonstrate how a mineralogical data base, for mill samples ranging from a few tens to a few hundred ppb PGE, may be used to determine metal balances during treatment of the Cu-Ni ores, and to indicate research areas to improve metal recoveries.

INTRODUCTION

Geology of the Sudbury Area Deposits

The discovery of Cu-Ni sulphides in the Sudbury, Ontario, area in 1883 soon led to their exploitation, and mining has been essentially continuous to the present day. These deposits (Fig. 1) are associated with sublayer intrusions along the margin of, and as outward radiating offset dykes from, an elliptical basin-shaped stratiform complex whose geology is now well-known (Pattison, 1979). Excellent summaries of the geology of the Sudbury Irruptive are given by Souch and co-workers (1969) and by Naldrett, Greenman and Hewins (1972), among others. The origin and emplacement of the Cu-Ni sulphide deposits of the Sudbury sublayer have been the subject of conflicting opinions and are reviewed briefly by Pattison (1979) who proposed that the sublayer mineralization had its origin from intrusion of a sulphide-rich melt formed either as the result of sulphide segregation of an impact-triggered magma or from previous concentrations in basic magmatic country rocks.

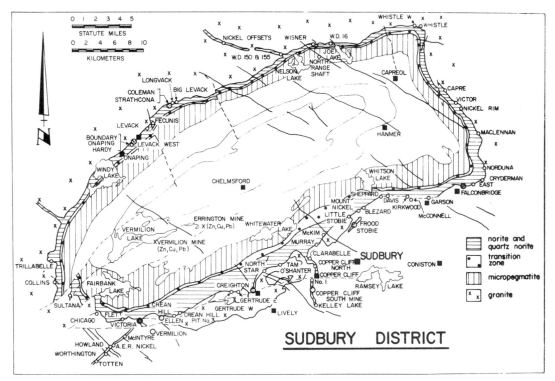

Fig. 1. Geology of the Sudbury district and location of
         mines (circles). Large filled circles indicate
         mines from which samples were studied by Cabri
         and Laflamme (1976). (Figure reproduced by
         permission of *Economic Geology*.)

Platinum-Group Mineralogy

The Cu-Ni deposits of the Sudbury area have been known to be an important source
of platinum-group elements (PGE) since the characterization of sperrylite ($PtAs_2$),
an important platinum-group mineral (PGM), by Wells (1889), just a few years after
discovery of the ore deposits. These deposits have been the principal source of
Canada's total PGE production which, in recent years, has ranked third in the
world. The determination of the Sudbury area PGM, and, indeed an understanding of
the distribution of the PGE has not kept pace with studies on the origin,
petrography, and major sulphide minerals of the deposits.

Detailed mineralogical studies are directly applicable to improving the recovery
of the PGE. It is essential to determine which minerals contain the elements of
economic importance. Not only is the identity, shape, size, and associations of
those minerals important but it must also be established whether any of the PGE,
and in what proportion, occur as dilute solid solutions in other minerals and in
what concentrations.

Hawley's (1962) comprehensive study of the mineralogy of these Cu-Ni deposits
reported only three known PGM: sperrylite, froodite ($PdBi_2$) and michenerite. The
composition of michenerite was not known precisely and the original samples were
lost. Cabri, Harris and Gait (1973) confirmed the presence of michenerite and

froodite from the Sudbury deposits and redetermined michenerite to be PdBiTe. Detailed mineralogical studies by Cabri and Laflamme (1976) of specially selected samples from nine Sudbury area mines and deposits revealed a total of thirteen PGM: froodite, insizwaite ($PtBi_2$), kotulskite (PdTe), merenskyite ($PdTe_2$), mertieite II [$Pd_8(Sb,As)_3$], michenerite, moncheite ($PtTe_2$), niggliite (PtSn), sperrylite, sudburyite (PdSb), and three unnamed minerals. Michenerite was reported the most common Pd mineral overall; sperrylite was the most common Pt mineral in the South Range and Offset deposits but moncheite was the principal Pt mineral in the North Range deposits where sperrylite was very rare.

The presence of PGE as solid solutions in more common minerals has been the source of some speculation, and attempts have been made to determine this more precisely. Hawley (1962) suspected that some of the PGE were present to some extent as a solid solution in major ore minerals. Rucklidge (1969) reported Pd in melonite ($NiTe_2$), a relatively rare mineral, from the Strathcona and Falconbridge deposits. Keays and Crocket (1970) stated that "Arsenic exhibits good positive correlation with Pd" but did not discuss these findings further. Cabri (1974) reported the first *in situ* analyses proving the occurrence of Pd and Pt as solid solutions in cobaltite from the Kanichee, Ontario, Cu-Ni deposit. Cabri and Laflamme (1976) showed that Pd, Pt and Rh occurred as solid solutions in Sudbury area arsenides and sulpharsenides and concluded "It therefore becomes important to ensure that beneficiation processes recover the arsenides and sulpharsenides for subsequent treatment". They did not think it likely that PGE occurred as dilute solid solutions in major ore minerals though they did not exclude that possibility in view of the large tonnages of ore processed. Chyi and Crocket (1976), in their study of partitioning of Pt, Pd, Ir and Au in minerals from the deep ore zone at the Strathcona mine, concluded that Pd was preferentially associated with pentlandite and Pt with chalcopyrite to the extent of 1 to 1.5 ppm total Pt plus Pd. Chyi and Crocket also pointed out, however, that although preferential associations of Pd and Pt with specific sulphides occurred, it is difficult to establish whether a solid solution relationship is involved or whether these rather low concentrations may reside as discrete PGM.

Platinum-Group Element Content of the Ores

Published data on the PGE contents of Sudbury ores are sparse. Hawley, Lewis and Wark (1951), Hawley and Rimsaite (1953) and Hawley (1962) enumerated the earliest data available, and production figures (0.78 ppm total PGE) have to be used when average estimates are required for the district (Naldrett and Cabri 1976). Hoffman and others (1979) report detailed analyses of 72 samples for the six PGE as well as Au, Cu, Ni, Fe, Co, S, Zn, and Pb but stress that an unweighted average of the analyses is *not* representative of the grade of the mines studied.

Recovery of the Platinum-Group Elements from the Ores

An excellent review of the milling and extractive techniques employed on the Sudbury area ores has been published (Schabas, 1977). This publication includes a flow chart of Inco Metals Company's Sudbury operations concerned with the recovery of 15 elements: Ni, Cu, Fe, S, Au, Ag, Co, Se, Te, Pt, Pd, Ir, Rh, Ru, and Os. Most of the ore from the South Range mines is initially milled at the new Clarabelle Mill while ore from the Frood-Stobie mines are initially milled in the older Frood-Stobie Mill. The Ni-Cu bulk concentrate containing the bulk of the PGE from these two primary mills is fed into the secondary Copper Cliff Mill which produces three concentrates: Ni concentrate, Cu concentrate and pyrrhotite concentrate. The PGE from the South Range mines are therefore distributed into six principal products: Ni concentrate, Cu concentrate, Frood-Stobie Mill tailings, Clarabelle Mill tailings, Copper Cliff Mill tailings and pyrrhotite concentrate. It may be estimated that the initial Cu concentrate contains most of the Ag and about half the Pd and Au; the Ni concentrate holds the bulk of the remaining PGE as well as the remaining Pd and Au. A $Cu_{2-x}S$ concentrate, formed from the Ni concentrate, collects the Ag and Au values, and is added to the copper converter. The bulk of the PGE report in PGE-rich residues from the nickel and copper refineries and are shipped to Acton, U.K., for refining. A detailed mineralogical study on selected mill products, based on the earlier work on the ore samples, was initiated with a view to pin point more closely those areas requiring further research to increase PGE recoveries.

SAMPLE PREPARATION AND ANALYTICAL METHODS

Samples of mill products, supplied by Inco Metals Company, were split with a seven-compartment riffle into fractions as illustrated in Fig. 2. Smaller splits were effected by quartering. Approximately one quarter (or half) of each sample was wet-sieved into +100, -100+150, -150+200, -200+325 and -325 mesh fractions; the fractions coarser than 325 mesh were concentrated using gravity and magnetic separation methods described by Cabri and Laflamme (1976). Some difficulties were experienced with incomplete separations involving the magnetic fractions because of clumping of the fine-grained portions. The heaviest fractions were mounted as monolayers in cold-setting araldite, prepared as polished sections and examined by ore microscopy. A Quantimet Image Analyser was used to determine the proportions of the major sulphide minerals.

The effectiveness of the separation and grain mounting techniques has been studied using doped mixtures of sulphide minerals (Cabri and Laflamme, in preparation). Physically purified and sized pyrrhotite, chalcopyrite and pentlandite were mixed in the proportions of 50%, 25% and 25%, by weight, respectively, to simulate a sulphide concentrate of Sudbury ore. A predetermined number of sized sperrylite grains were added and the mixtures were treated to the same mineral separation and grain mounting techniques referred to above. The results indicate excellent recoveries down to about 270 mesh and poorer recoveries for fractions finer than 270 mesh. The results for a typical test are;
-100+140 mesh: 100% recovery of sperrylite grains in 1st sink fraction.
-140+200 mesh: 100% recovery of sperrylite grains distributed as 84% in 1st sink, 8% each in 2nd and 3rd sink fractions.
-200+270 mesh: 73% recovery of sperrylite grains distributed as 63% in 1st sink, 7% in 2nd sink, nil in 3rd sink, and 3% in 4th sink fraction.
-270+325 mesh: 20% recovery of sperrylite grains distributed as 10% in 1st sink and 3% each in the 2nd, 3rd and 4th sink fractions.
All PGM were identified by the electron microprobe using energy dispersive semi-quantitative analysis (EDX); wavelength dispersive techniques (WDX) were used for quantitative analyses, as required. The techniques described by Cabri and Laflamme (1976) were used for trace-level PGE analyses. A scanning electron microscope (SEM) was used to resolve fine-grained inclusions and PGE element zoning in sulpharsenides by back-scattered electron (BSE) techniques.

Specific cuts of the samples were analysed in duplicate by the Central Laboratory, Inco Metals Company, at Copper Cliff (Table 1). Nickel and copper were determined by standard acid dissolution with atomic absorption finish. Arsenic analyses were performed by chloride evolution and a colourimetric finish. Pt, Pd and Au were determined by fire assay using lead collection cupellation with 100 mg Ag as the carrier. The resultant beads were then sparked on an emission spectrometer. Ir, Rh, and Ru were also determined by lead collection fire assay but were cupellated with 10 mg Pt as carrier and the resultant bead arced with a graphite anode and a copper rod cathode using a DC arc on an emission spectrometer. The Ru was

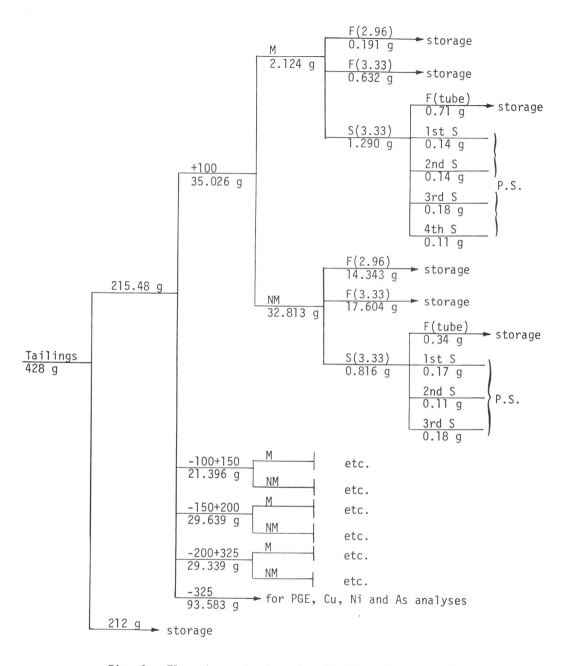

Fig. 2. Flow sheet showing distribution of some of the
         fractions and their treatment for a sample of
         mill tailings. F(2.96), S(2.96) are for float
         and sink fractions at 2.96 S.G.; M, NM are for
         magnetic and non-magnetic fractions; P.S. =
         polished sections.

Louis J. Cabri

corrected for furnace losses using a radioactive tracer.

TABLE 1   Platinum-Group Elements, Arsenic, Nickel
and Copper Analyses for a Mill Tailing

Element	Whole Sample (206 g)	+325 Mesh** (138.767 g)	Sample Distri- bution %	-325 Mesh (64.7 g)	Sample Distri- bution %	Loss*** (2.533 g)	Sample Distri- bution %
Pt ppb*	641	469	49.3	995	48.8	995	1.9
Pd ppb	353	330	63.0	401	35.6	401	1.4
Rh ppb	127	135	71.6	110	27.2	110	1.1
Ru ppb	89	86	65.1	96	33.6	96	1.3
Ir ppb	48	48	67.4	48	31.4	48	1.2
As %	0.0058	0.00507	58.9	0.0073	39.5	0.0073	1.5
Ni %	0.705	0.671	64.1	0.774	34.5	0.774	1.4
Cu %	0.205	0.199	65.3	0.218	33.4	0.218	1.3
Distribution %	100	67.36	-	31.41	-	1.23	-

*1 ppb = 1×10^{-9} by weight; **calculated; ***calculated, values estimated to be the same as for -325 mesh fraction. It should be noted that these values relate only to the sample studied and are *not* representative of any specific tailings.

RESULTS

General Mineralogy

A total of 69 monolayer polished sections were prepared from 20 sink-float products for three mill products as outlined in the example shown in Fig. 2. Careful microscopic examination confirmed that the principal ore minerals are pyrrhotite, chalcopyrite and pentlandite with lesser amounts of cobaltite/ gersdorffite, pyrite, marcasite, millerite, cubanite, violarite, galena, sphalerite, magnetite, ilmenite, mackinawite and argentopentlandite. Rare constituents were undefined Bi-tellurides, hessite, altaite, electrum, arseno- pyrite, maucherite, nickeline, molybdenite, parkerite and 8 PGM: sperrylite, michenerite, froodite, merenskyite, moncheite, sudburyite, hollingworthite, and irarsite in approximately the relative abundance reported for ore samples by Cabri and Laflamme (1976) except for hollingworthite and irarsite whose presence in Sudbury ores had not been previously documented.

Earlier work on ore samples by Cabri and Laflamme (1976) had demonstrated that Pd, Pt and Rh could not be detected in the major sulphide minerals by *in situ* microprobe analyses at minimum detection limits (MDL) of 0.03, 0.04 and 0.05%, respectively, for chalcopyrite, pyrrhotite,[2] pentlandite, magnetite, and millerite. It was therefore felt unnecessary to perform the tedious low-level analyses for Pd, Pt and Rh on the major sulphides in spite of significant and measurable amounts of Pd detected in pentlandite in other ores such as Lac des Iles, Ontario, and the Stillwater Complex, Montana (Cabri and Laflamme, 1979, Conn 1979). On the other hand, it has already been demonstrated that sulpharsenides such as cobaltite and gersdorffite contain measurable quantities of Pt, Pd and Rh

---

[2]MDL for pyrrhotite was 0.05, 0.04 and 0.03% for Pd, Pt, and Rh, respectively.

in solid solution (Cabri, 1974, Cabri and Laflamme, 1976). Sulpharsenides were therefore analysed.

Pentlandite and Pyrrhotite

Quantitative electron microprobe analyses were made on pentlandite and pyrrhotite in order to determine the nickel content for use in metal balance calculations. The contribution of nickel-bearing minerals such as violarite, millerite, argento-pentlandite and gersdorffite is not taken into account due to their negligible quantity. Ten randomly selected grains of pentlandite had a mean value of 34.3% Ni with a range from 33.2 to 34.8%.

Nickel is known to occur as a solid solution in both the magnetic monoclinic variety of pyrrhotite [$(Fe,Ni)_7S_8$] and in the hexagonal variety [$(Fe,Ni)_9S_{10}$]. Because the monoclinic variety is more abundant, 17 randomly selected monoclinic pyrrhotite grains were analysed in contrast to only 8 grains of hexagonal pyrrhotite. The mean values and range for the monoclinic and hexagonal pyrrhotites are, respectively: 0.64% Ni (0.37-1.04) and 0.625% Ni (0.36-0.81). For metal balance calculation purposes a Ni content of 0.64% will be applied for all the pyrrhotite.

The Sulpharsenides

Cobaltite (CoAsS) and gersdorffite (NiAsS) exhibit complete solid solution, usually with some replacement of Co-Ni by Fe. Cabri and Laflamme (1976) showed that these sulpharsenides are typically non-homogeneous in samples from the Sudbury area. Quantitative analyses to differentiate between these minerals were not done and, for convenience, all calculations are made on the basis of cobaltite.

The size of these sulpharsenide grains in mill sample A ranged from less than 5 microns to about 80 X 145 microns, as measured in polished sections. The liberation characteristics for 108 sulpharsenide grains in that sample were: 52% free grains, 33% included in pyrrhotite, and 15% attached to pyrrhotite-chalcopyrite-gangue. The sulpharsenides are often euhedral, especially the smaller unbroken grains (Figs. 3 and 4), and these are the ones usually found included in pyrrhotite. A total of 243 spot analyses on 50 randomly selected grains were made to determine their Pd, Pt and Rh contents. The results are given in Table 2 and the MDL for Pd, Pt and Rh is 0.03, 0.05 and 0.03% (or 0.05%), respectively.

Cobaltite/gersdorffite grains indicating Rh or Ir values with the EDX were also examined with a SEM at magnifications up to 4,000X to determine whether small PGM inclusions were present to account for the PGE values. The large difference in atomic numbers between the PGE and the transition metal sulpharsenides produces high visibility of the higher atomic number elements in the back-scattered electron mode. Thus, not only are submicron PGM inclusions visible but also zones with diffuse, high PGE values are rendered visible (Figs. 4 and 5). This study has shown, therefore, that PGE values in Co-Ni sulpharsenides may be due to PGM sulpharsenide inclusions; to a dilute solid solution of PGE; to a diffuse, but more concentrated, solid solution zoning of PGE in the Co-Ni sulpharsenide; or to combinations of these situations. A Rh-rich gersdorffite grain amenable to a complete quantitative analysis gave:

$(Ni_{0.38}Co_{0.37}Fe_{0.20}Rh_{0.06}Pt_{<0.01}Pd_{<0.01})As_{1.03}S_{0.95}$. The average PGE content of

cobaltite/gersdorffite for two mill samples was calculated for all the grains

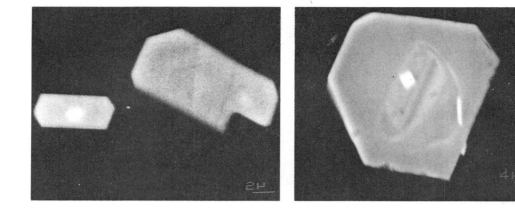

Fig. 3. Back-scattered electron photo-   Fig. 4. Back-scattered electron photo-
         micrograph of two cobaltite/   micrograph of a euhedral
         gersdorffite inclusions in   cobaltite/gersdorffite
         pyrrhotite. White area is a   inclusion in pyrrhotite with
         hollingworthite (RhAsS)   a cube and a lamellae of
         inclusion.   irarsite (IrAsS). Faint white
                                                        areas are due to high Ir and
                                                        Rh concentrations in cobaltite.

TABLE 2   Distribution and Content of Pd, Pt,
and Rh in Cobaltite/Gersdorffite

				Ranges in Wt. %			
	<0.03	0.03-0.10	<0.05	0.05-0.10	0.10-0.25	0.25-0.50	>0.50
Sample A*							
Pd %	37	42	-	-	17	4(0.32)[†]	-
Pt %	-	-	94	1	4	1(0.29)	-
Rh %	88	5	-	-	2	1	4(8.1)
Sample B**							
Pd %	2	40	-	-	20	29	9(0.56)
Pt %	-	-	71	16	13(0.24)	-	-
Rh %	-	-	71	4	7	4	13(2.96)
Sample C***							
Pd %	9	35	-	-	44	5	7(1.04)
Pt %	-	-	94	6(0.08)	-	-	-
Rh %	81	9	-	-	6	4(0.38)	-

*   144 spot analyses on 29 randomly selected grains.
**   45 spot analyses on 10 randomly selected grains.
***   54 spot analyses on 11 randomly selected grains.
†   Highest values in parentheses.

analysed with the electronprobe for which no PGM inclusions were detected with the
SEM. Therefore, the grains with high Rh-Ir zones surrounding PGM inclusions
(Figs. 4 and 5) were also not included in the averaging calculations. It is

important to discuss the method of calculation in some detail. It should also be noted that the MDL selected is based on the conservative premise that there is only a 0.03% probability that background measurements deviate by more than 3σ (Cabri and Laflamme, 1976). On the other hand, it is unrealistic to expect that the grains with "nil" PGE values, i.e., less than the conservative MDL selected, do not actually contain some lesser amounts of PGE, between 0.000% and the MDL.

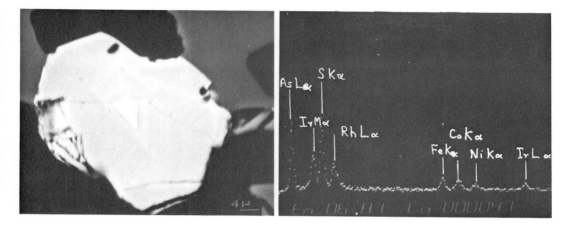

Fig. 5. Back-scattered electron photo-micrograph of an irarsite inclusion (white) in a euhedral cobaltite grain, itself included in pyrrhotite. Note that the crystal faces of irarsite are parallel to cobaltite's as are the diffuse Ir-Rh rich zones around the irarsite.

Fig. 6. EDX spectra of a hollingworthite-irarsite inclusion showing peaks due to As, Ir, S, Rh, as well as minor Fe, Co, and Ni from the surrounding cobaltite/gersdorffite.

The mass of each cobaltite/gersdorffite grain analysed was calculated using the area measured from polished sections, multiplied by 1/2 (length + width), to obtain the volume. The density of cobaltite was taken as 6.3 g cm^{-3}. Each cobaltite/gersdorffite grain's mass was then multiplied by the mean of the PGE content, obtained from 4-10 random spot analyses, to calculate its PGE content. The average PGE content in the cobaltite can then be calculated and this was done in two ways. The first was to include, in averaging calculations, all grains with "nil" PGE (due to the MDL selected) as zero values and the second was to assign a value of 1/3 MDL for those grains (Table 3).

The Platinum-Group Minerals

A total of eight PGM species was found in the mill samples: sperrylite ($PtAs_2$), michenerite (PdBiTe), froodite ($PdBi_2$), merenskyite ($PdTe_2$), hollingworthite (RhAsS), sudburyite (PdSb), irarsite (IrAsS) and moncheite ($PtTe_2$) in approximately the relative abundance previously reported for ore samples (Cabri and Laflamme, 1976). Quantitative analyses of selected PGM were also consistent with the PGM from the ore samples. Sperrylite, hollingworthite and irarsite will be discussed in more detail as they are of concern in calculating metal balances for

As, Rh, Pd and Pt.

TABLE 3 Calculated Pt, Pd and Rh Contents of Cobaltite (ppm)

	Mill Sample A		Mill Sample B	
	Using Nil Values	Using 1/3 MDL	Using Nil Values	Using 1/3 MDL
Av. Pt content of cobaltite in ppm	56.9	221.5	170	320
Av. Pd content of cobaltite in ppm	610	643	1840	1840
Av. Rh content of cobaltite in ppm	300	398	670	820

These results will be applied to the metal balance calculations below.

A total of nine *sperrylite* grains, ranging in cross section from 15X15 to 55X55 microns, was found in mill sample A. Three grains were free (33%), three were attached to cobaltite/gersdorffite (33%), two were attached or included in pyrrhotite (22%) and one grain was attached to galena and magnetite (11%). *Hollingworthite* and *irarsite* have not been reported from Sudbury ores previously. The grains found were typically euhedral and all occurred as minute inclusions in cobaltite/gersdorffite 1X4 to 3X3 microns in cross section. The grains were confirmed by EDX only as they were too small for quantitative analysis. Most EDX spectra had both Rh and Ir peaks (Fig. 6). It was determined that both end-members occurred though most grains were variaties, with iridian hollingworthite occurring more frequently than rhodian irarsite.

METAL BALANCES

The dependance of metal balance calculations on a detailed mineralogical investigation will be demonstrated for the PGE associated with sulpharsenides for mill sample A. The detailed mineralogical study has revealed that the Co-Ni sulpharsenides are the major contributors to the As value of the sample, probably in excess of 99%. Minerals such as sperrylite, maucherite, nickeline, and arsenopyrite are too sparce to contribute much to the As value. Knowing the As value in the original sample and in the -325 mesh fraction (Table 1) it is possible to calculate the As content of the +325 mesh fraction on which the mineralogical work was done, assuming that the weight loss of 1.2% in sample treatment had the same As value as the -325 mesh fraction. The mass of the sperrylite grains found, calculated from the area of each grain in the polished section as was done for cobaltite (above), was used to determine the As content due to sperrylite. This value comes to only 0.039% of the total As content - a rather insignificant quantity. Even if 50% of the sperrylite grains present were not found in the polished sections (which is highly unlikely from the tests done with doped samples) the sperrylite contribution to the As content would still be less than 0.1%. The calculated As content of the +325 mesh fraction, less the very small amount of As due to sperrylite, was then used to calculate the weight of cobaltite in this fraction. The weight of cobaltite multiplied by the average Pt, Pd and Rh contents of cobaltite (Table 3) divided by the sample weight (138.767g) gives the Pt, Pd and Rh values of the +325 mesh fraction due to solid solution in sulpharsenides. The percentage of the total Pt, Pd and Rh in this fraction may then be calculated using "nil" values or 1/3 MDL values (Table 3) as shown in Table 4.

Assuming that all the -325 mesh material has sulpharsenides with the same relationships and PGE concentrations as the +325 mesh fraction, it is possible to calculate the percentage of Pt, Pd, and Rh in solid solution in sulpharsenides (Table 5) for that fraction.

TABLE 4   Percentage Pt, Pd and Rh in Solid
          Solution in Sulpharsenides for the
          +325 Mesh Fraction (Mill Sample A)

Element	Using Nil Values	Using 1/3 MDL
Pt	1%	5%
Pd	21%	22%
Rh	25%	33%

TABLE 5   Percentage Pt, Pd and Rh in Solid
          Solution in Sulpharsenides for the
          -325 Mesh Fraction (Mill Sample A)

Element	Using Nil Values	Using 1/3 MDL
Pt	0.9%	3.6%
Pd	25%	26%
Rh	44%	58%

Therefore, for the total sample, the percentages that occur as solid solutions in cobaltite/gersdorffite are: Pt 1%, Pd 22.3%, and Rh 31.2% using the most conservative assumptions. It is likely, however, that the percentages are closer to: Pt 4.5%, Pd 23.3% and Rh 41.2% or higher, depending on the true PGE concentrations below the selected MDL.

CONCLUSIONS AND RECOMMENDATIONS

It has been possible to determine, by a detailed mineralogical investigation, the nature of some of the PGE in mill samples containing very low levels of these elements (from a few tens to a few hundred ppb). Pt, Pd, Rh and Ir have been found in discrete PGM but the exact proportion present is still being evaluated. On the other hand, the presence of significant quantities of Pd and Rh as a solid solution in sulpharsenides makes it worthwhile to attempt flotation research towards improving their recovery. Sperrylite may have similar flotation proper- ties to the sulpharsenides because of similar chemistry and crystal structure, and, because it was found to occur mainly as liberated grains or attached to sulpharsenides (66%), improved sulpharsenide recovery is therefore likely to improve Pt recovery as well. Likewise, the close association of sperrylite and sulpharsenides with pyrrhotite suggests that research into better separation of these minerals is attractive.

ACKNOWLEDGEMENTS

The author acknowledges Inco Metals Company for encouragement in this study, for providing the samples and for the chemical analyses. The author is grateful to Dr. D.C. Harris for the SEM studies and for technical discussions and for the outstanding technical support provided by Messrs. J.H.G. Laflamme, Y. Bourgoin, P. Carrière, M. Beaulne and R.G. Pinard. The author has also benefited from critical comments by Dr. J.E. Dutrizac.

REFERENCES

Cabri, L.J. (1974). The significance of detailed mineralogical studies of sulfide ores. *Geol. Soc. Amer., Abs. with Programs, 6,* No. 7, 677.

Cabri, L.J., D.C. Harris, and R.I. Gait (1973). Michenerite (PdBiTe) and froodite (PdBi$_2$) confirmed from the Sudbury area. *Can. Mineral., 11,* 903-912.

Cabri, L.J., and J.H.G. Laflamme (1976). The mineralogy of the platinum-group elements from some copper-nickel deposits of the Sudbury area. *Econ. Geol., 71,* 1159-1195.

Cabri, L.J., and J.H.G. Laflamme (1979). Mineralogy of samples from the Lac des Iles area, Ontario. CANMET Report 79-27, Dept. Energy, Mines and Resources, Ottawa.

Chyi, L.L., and J.H. Crocket (1976). Partition of platinum, palladium, iridium and gold among coexisting minerals from the deep ore zone, Strathcona mine, Sudbury, Ontario. *Econ. Geol., 71,* 1196-1205.

Conn, H.K. (1979). The Johns-Manville platinum-palladium prospect, Stillwater complex, Montana, U.S.A. *Can. Mineral., 17,* 463-468.

Hawley, J.E. (1962). The Sudbury ores: their mineralogy and origin. *Can. Mineral., 7,* 1-207.

Hawley, J.E., C.L. Lewis, and W.J. Wark (1951). Spectrographic study of platinum and palladium in common sulfides and arsenides of the Sudbury district, Ontario. *Econ. Geol., 46,* 149-162.

Hawley, J.E., and Y. Rimsaite (1953). Platinum metals in some Canadian uranium and sulfide ores. *Amer. Mineral., 38,* 463-475.

Hoffman, E.L., A.J. Naldrett, R.A. Alcock, and R.G.V. Hancock (1979). The noble metal content of ore in the Levack West and Little Stobie mines, Ontario. *Can. Mineral., 17,* 437-451.

Keays, R.R., and J.H. Crocket (1970). A study of precious metals in the Sudbury nickel irruptive ores. *Econ. Geol., 65,* 438-450.

Naldrett, A.J., and L.J. Cabri (1976). Ultramafic and related mafic rocks: their classification and genesis with special reference to the concentration of nickel sulfides and platinum-group elements. *Econ. Geol., 71,* 1131-1158.

Naldrett, A.J., L. Greenman, and R.H. Hewins (1972). The main irruptive and sub-layer at Sudbury, Ontario. *Internat. Geol. Congr. 24th, 4,* 206-214.

Pattison, E.F. (1979). The Sudbury sublayer. *Can. Mineral. 17,* 257-274.

Rucklidge, J. (1969). Electron microprobe investigations of platinum metal minerals from Ontario. *Can. Mineral., 9,* 617-628.

Schabas, W. (1977). The Sudbury operations of Inco Metals Company. *Can. Min. J., 98,* No. 5, 10-90.

Souch, B.E., T. Podolsky, and staff (1969). The sulfide ores of Sudbury: their particular relationship to a distinctive inclusion-bearing species of the Nickel Irruptive. *Econ. Geol., Monog. 4,* 252-261.

Wells, H.L. (1889). Sperrylite, a new mineral. *Am. J. Sci., 37,* 67-70.

Silver Mining in Peru

G. Florez Pinedo

Mining Engineer, Executive President, Empresa Minera
del Centro del Peru, CENTROMIN PERU, Lima-Peru.

ABSTRACT

With 43 million ounces per year (26 millions refined), Peru stands in the third position within the world silver production.

The reserves come up to 610 million ounces, with potential superior to the 1,200 million (that is more than 8% of the world's silver-bearing resources).

In Peru silver is found throughout the Cordillera de los Andes, and generally in veins accompanying copper and lead in the form of polibasite, argentite, stephanite, pyrargyrite, native silver and lead-silver sulphides.

It constitutes an important item in Peru's international commerce, contributing more than 10% of the exportations value. Approximately 60% of this metal is refined by CENTROMIN PERU.

It is not possible to conjecture important increases in production, except for a higher extraction volume which would be capable of covering full installed capacity reaching 45 million ounces per year. (29 Millions belonging to CENTROMIN PERU.)

This presently requires some technological adjustments in CENTROMIN PERU's metallurgical complex.

KEYWORDS

Geology; mineral deposits; reserves and resources; production and selling; perspectives on non-conventional silver ores and other resources.

INTRODUCTION

Peru, a country with a mining tradition since the pre-Inca time, has produced silver in important quantities which were incremented in the colonial period during the XVI century.

The first statistic figures show that already during the first part of the XIX century two and a half million ounces were being produced, these came from rich deposits of native silver.

Later, when the construction of the Peruvian Central Railroad was concluded, the development of argentiferous mining industry was promted, treating not only native silver deposits but also in an increasing manner silver was obtained from polimetallic deposits especially from lead and copper, thus at the initiation of the XX century Peru already produced nearly seven million ounces per year, exceeded 15 million towards 1950 to reach a level above 43 million ounces during 1979, thus maintaining a privileged place amongst the silver producers in the world market.

This metal consitutes an item of importance in Peru's international commerce, having reached 320 million dollars of export value in 1979 and constituting generally a proportion in excess of 10% with respect to the total foreign trade revenue which comes into the country, notwithstanding its growing utlization in the local market.

GEOLOGY AND RESERVES

Silver in Peru occurs with greater frequency as a by-product in copper and lead ores and also, in a good number of cases, as strictly argentiferous deposits.

As a main product it is generally found in veins associated at tertiary volcanic rocks in the Andean Cordillera on the occidental side and at the central and southern part of the country.

The principal silver ores are: polibasite, argentite, stephanite and silver and lead sulphosalts.

As a by-product it is generally found in copper, lead and zinc deposits, also over the Andean Cordillera, generally in veins and replacement bodies.

Figure 1 shows the argentiferous zones with the greatest production, outstanding as argentiferous departments, we have Pasco, Lima, Huancavelica, Ayacucho, Arequipa and Puno.

The argentiferous reserve proven in Peru, totals 610 million ounces, equivalent to 10% of the world wide reserves, and the potential reserve reaches 1,200 million ounces which totals a volume of argentiferous resources of 1,810 million ounces equivalent to 8% of world production.

PRODUCTION

CENTROMIN PERU's silver production as refined metal is over 60% of Peru's production. In 1979 CENTROMIN's production was 25.5 million ounces. The remaining silver is found in copper and lead concentrates.

CENTROMIN PERU is one of the two largest producers in the world. Among the main Peruvian silver producers are: Buenaventura, Milpo, Atacocha, Huaron and Millotingo. Southern Peru Copper Corporation is also an important producer of silver, since in the mines of Toquepala and Cuajone silver is found in copper concentrates. The blister of these two mines is treated by the State owned company, MINERO PERU, in its refinery at Ilo. (See Table 1).

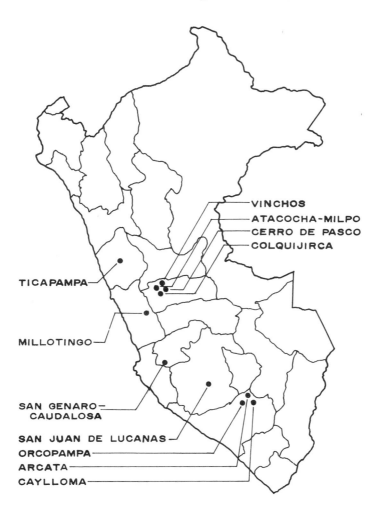

Fig. 1   Main Silver Producers in Peru

Silver price prior to 1979 was not attractive enough to encourage the search of new methods for the extraction of this metal from complex ores and other unconventional sources (treatment of tailings, films, argentiferous pyrites and ferrites). That is why production has been more or less static.

COMMERCIALIZATION

The most important markets for Peruvian refined silver are the U.S.A., European Economic Community and Japan.

It is worth mentioning that local market consumption has been increasing lately in the manufacture of silver ware, jewelry and silver plating. In 1979 four million ounces of refined silver were sold in the local market.

G. Florez Pinedo

TABLE 1   Silver Production from Peruvian Mines, 1979

Companies	Silver Content Million Ounces
CENTROMIN PERU	12' 810
BUENAVENTURA	4' 336
SOUTHERN PERU C.C.	2' 412
ARCATA	2' 117
ALIANZA	2' 025
MILPO	1' 587
HUARON	1' 580
CASTROVIRREYNA, CIA.	1' 538
CORP. CASTROVIRREYNA	1' 344
NORTHERN PERU	1' 248
ATACOCHA	1' 234
MILLOTINGO	1' 136
RAURA	1' 024
OTROS	9' 300
TOTAL	43' 691

PERSPECTIVES

Production plans on the basis of conventional sources for the future do not present important increases in volumes because silver is constituted mainly a by-product in mid scale underground mines. However, it is sure that our country will maintain a privileged place amongst the producers of silver, in as much as our concentrates are similar to those of other large world producers.

An important improvement in price levels will encourage silver producers to increase their production. In the case of Peru for example, some companies that recover metals from old tailings are effectuating researches orientated towards silver recuperation.

Likewise, CENTROMIN PERU is carrying out investigations to find methods which, under better price conditions, can be economically convenient. Within this line of thought the treatment of silver complex ores (silver argentiferous pyrites, silver and lead sulphosalts) as well as old silver bearing tailings have been considered.

Other work is being carried out to recover this metal from the treatment of ferrites (zinc slimes) which are complex zinc-iron oxides (14-16 ounces of silver per DST). These ferrites are the tailings of zinc refinery we have in La Oroya and are produced at a level of 66 thousand DST per year, having accumulated more than a million tons. In this way CENTROMIN PERU expects to raise its production from 25.5 to 30 million ounces of silver per year.

The remarkable price increase silver has reached during the last months, up to March more or less, and the strong descent appreciated in April and May shows a chaotic tendency that does not alllow the coherent outlining of plans at least in the short and medium term.

A mutual understanding between producers and consumers could achieve the objective of fair prices and minimize significant fluctuations.

In Peru there is the idea of creating, in an immediate term, a promotional mining policy that would permit, particularly in the case of small and medium mining companies, a quick development which would result in a significant increase in argentiferous production.

Treatment of Complex Silver Arsenide Concentrate in Nitric Acid System

W. Kunda*

* Research Consultant, Sherritt Research Centre,
Sherritt Gordon Mines Limited, Fort Saskatchewan,
Alberta, Canada, T8L 2P2

ABSTRACT

A complex silver arsenide-sulphide concentrate containing bismuth, cobalt, nickel, copper, zinc, lead and iron as major constituents, was leached in nitric acid. The leach solution, which contained 95% to 99% of the Ag, Bi, Co, Ni, Cu, As and Fe present in the ore, was further processed to separate the metals into the following fractions: (1) silver chloride, (2) bismuth oxychloride and/or hydroxide, (3) arsenic-iron precipitate and (4) Ni-Co-Cu-Zn sulphide.

Silver was precipitated from solution with sodium chloride. Bismuth oxychloride-hydroxide and arsenic-iron were precipitated by controlled ammonia addition and cobalt-nickel-copper-zinc sulphides were precipitated with ammonium sulphide.

Silver chloride was further processed to a high purity silver powder by heat treatment with sodium carbonate.

The parameters controlling each processing step are discussed.

KEYWORDS

Silver concentrate; nitric acid leach; arsenide; sulphide; silver chloride; calcining; silver powder; bismuth residue; iron-arsenic residue; nickel-cobalt-zinc sulphide; metals separation.

INTRODUCTION

Statistics published by The Silver Institute[1] shows that in 1979 the United States produced 1,856,240 kg silver from primary sources. This represents only one-third of the silver consumption in the U.S. Production in Canada in 1978[2] was only 1,200,000 kg. Generally, consumption exceeds production by a factor of two.

At the present time, the main sources of silver are copper, zinc and lead based ores. Only a small fraction of silver is derived from silver ore. Canada has large deposits of silver ore in Ontario and the Northwest Territories. However,

both these ores are mineralogically very complex and contain a large quantity
of desirable and undesirable impurities like Ni, Co, Cu, Zn, Bi, Pb, As and Fe.

Silver residues derived from processing base metals are usually treated by
hydrometallurgical methods using cyanide[3], thiosulphate[4], thiurea[5],
chloride[6,7], sulphate[8], or nitrate[9] systems. Native silver is also
recovered by amalgamation treatment. The silver concentrate is processed
almost exclusively by pyrometallurgical methods. However, due to environmental
regulations, only low arsenic concentrates are accepted by the smelters. There
are also restrictions on the content of the other non-ferrous impurities. Mine
producers are penalized for such valuable metals as Ni, Co and Cu.

Processing of high arsenic silver ore from the Northwest Territories by the
conventional method is very difficult and only a small fraction of the mined ore
can be used by the smelters. Therefore, there was a great need for a process to
treat this type of concentrate and recover not only silver, but also Ni, Co, Cu
and possibly other metals as by-products.

A laboratory study was carried out on the treatment of such silver concentrate.
The leaching of this concentrate in nitric acid resulted in almost complete
extraction of all metals with the exception of lead which remained in the leach
residue. The leach solution was further processed with the recovery of (1) silver
chloride, (2) bismuth oxychloride or hydroxide, (3) arsenic-iron residue and (4)
Ni-Co-Cu-Zn sulphide residue. Silver chloride was further processed to a high
purity silver powder.

                    MATERIALS

Silver Concentrate

Two types of silver concentrates were used for the tests: (1) jig concentrate
and (2) flotation concentrate. The jig concentrate was much coarser and was
comprised of coarse fractions containing metallic silver and metallic bismuth.
The chemical composition of both concentrates varied widely and was dependent
on the mineralogical composition of the orebody.

A study was carried out on various samples of both concentrates and also on a
mixture of jig and flotation concentrates originated from various orebodies.
The chemical composition, screen analyses and mineralogical identifications of
species found in the concentrate are shown in Tables 1, 2 and 3 respectively.

Reagents

The process required the following reagents, all of which were commercial grade
except $(NH_4)_2S$ which was prepared in the laboratory:

       (1)   Nitric acid
       (2)   Ammonium hydroxide
       (3)   Sodium chloride
       (4)   Sodium carbonate
       (5) $(NH_4)_2S$

TABLE 1 Silver Ores Used for Testing

Chemical Analyses (%)

Type of Ore	Sample No.	Ag	Ni	Co	Cu	Zn	Bi	As	Fe	Pb	Sb	Ca	Mg	Al	Si	S	CO_2	U_3O_8
Jig Concentrate	1	8.33	1.98	1.41	0.59	0.16	7.90	6.74	9.03	2.11	0.12	5.0	2.66	4.73	18.4	5.16	2.31	n.a.
	2	8.16	1.74	1.64	1.06	0.62	2.74	7.75	12.10	2.70	n.a.	4.0	1.93	3.50	13.6	6.58	1.49	0.037
	3	31.50	0.68	0.56	0.75	1.67	1.01	2.61	5.43	9.96	n.a.	0.9	0.73	2.85	11.1	6.69	0.55	0.0006
Flotation Concentrate	1	3.04	0.94	1.03	6.42	1.06	0.91	4.40	18.30	6.05	n.a.	3.4	2.52	1.21	8.3	18.3	n.a.	n.a.
	2	2.70	0.88	1.15	7.47	2.48	0.88	4.28	17.10	5.90	n.a.	3.3	1.38	2.07	10.1	16.1	1.52	0.034
	3	0.83	0.19	0.24	2.33	10.80	0.19	0.63	7.25	30.00	n.a.	0.8	0.67	3.02	10.9	15.8	0.56	0.0012

W. Kunda

TABLE 2   Screen Analysis of Silver Ore Used for Testing

Tyler Screen Mesh	Jig Concentrate No.			Flotation Concentrate No.		
	1	2	3	1	2	3
+28	0.4	0	14.9	0.4	0	tr
28/65	8.6	tr	20.8	10.7	tr	2.2
65/100	23.2	0.1	8.2	15.5	0.1	5.6
100/150	34.0	2.2	8.4	17.4	4.1	11.1
150/200	38.0	4.9	6.9	6.5	4.8	13.3
200/250		1.1	1.0	4.8	0.8	1.9
250/325		9.7	5.6	11.0	5.6	12.7
−325		82.0	34.2	34.0	84.6	53.2

TABLE 3   Mineralogical Species Found in Silver Ore

Mineral	Chemical Formula
Chalcopyrite	$CuFeS_2$
Pyrite	FeS_2
Marcasite	FeS_2
Sphalerite	ZnS
Galena	PbS
Tetrahedrite	Cu_3SbS_3
Bismuthinite	Bi_2S_3
Native Ag	Ag
Native Bi	Bi
Pearceite	$(AgCu)_{16}(As,Sb)_2S_{11}$
Acanthite	Ag_2S
Niccolite	$NiAs$
Gersdorffite	$NiAsS$
Rammelsbergite	$NiAs_2$
Safflorite	$(Co, Fe)As_2$
Skutterudite	$(Co, Ni, Fe)As_3$
Arsenopyrite	$FeAsS$
Cobaltite	$CoAsS$
Matildite	$AgBiS_2$
Pavonite	$AgBi_3S_5$
Magnetite	Fe_3O_4
Siderite	$FeCO_3$

PROCEDURE AND EQUIPMENT

Leaching

Leaching tests were carried out in 2.5 L or 6.0 L operating capacity stainless steel autoclave, equipped with a mechanical agitator consisting of a shaft with a marine type impeller at the bottom and paddle type impeller above. Speed of agitation was 900 rpm. The autoclave was heated with a gas burner. An outside water spray was used for cooling. The gas burner was monitored by a temperature controller.

The autoclave was charged with a calculated quantity of concentrate and water, then sealed and heated to a predetermined reaction temperature. Oxygen was admitted at the reaction temperature and then a calculated quantity of 15.7 N nitric acid was slowly added within about 15 minutes. Due to a very exothermic reaction, cooling was necessary. The duration of the tests was 30 to 180 min. In some tests, samples were withdrawn during the run to monitor the progress of the leach.

On completion of the test, leach slurry was discharged and filtered and washed residues and solutions were analysed.

Separation of Metals

Metals extracted in the leach solution were separated into various fractions by the addition of: NaCl, ammonia and ammonium sulphide. All of these operations were carried out in an open vessel at temperatures of 25°C to 80°C.

Conversion of Silver Chloride to Silver Metal

Silver chloride was mixed with sodium carbonate and calcined in an air atmosphere at elevated temperature using a Lindberg furnace. The heat-treated material was washed with water in order to separate silver product from sodium chloride and sodium carbonate.

DISCUSSION AND RESULTS

Leaching

Silver concentrates used for study varied widely in chemical and mineralogical composition. Application of the hydrometallurgical method appeared to offer many advantages: higher recovery of Ag, recovery of other metals present in the ore and a safer disposal of environmentally undesirable products. Metals of interest in the concentrate, in order of their importance, were: Ag, Co, Ni, Cu, Bi, Zn and Pb. Selective recovery of desirable metals from this concentrate during leaching was unlikely to be achieved and, therefore, it was decided to choose a system which will dissolve all valuable metals, and then separate these metals by selective precipitation from leach solution.

All of the above mentioned metals form very soluble nitrate salts and therefore the nitric acid system was selected for leaching. Nitric acid reacts with native metals and metal sulphides according to the following reactions:

Metals:

$$2HNO_3 \rightleftharpoons NO_2 + NO + O_2 + H_2O \tag{1}$$
$$Me + 2NO_2 \longrightarrow MeNO_3 + NO \tag{2}$$
$$2NO + O_2 \longrightarrow 2NO_2 \tag{3}$$

Metal Sulphides:

$$3MeS + 4HNO_3 \longrightarrow 3MeNO_3 + 2H_2O + 3S° + NO \tag{4}$$
$$NO + 0.5\ O_2 \longrightarrow NO_2 \tag{5}$$
$$2NO_2 + H_2O \longrightarrow HNO_3 + HNO_2 \tag{6}$$

The reaction of nitric acid with metals proceeds very rapidly even at low temperature. Reaction products are metal nitrates and nitric oxide. Reactions involving metal sulphides and metal arsenides proceed much slower and require a breaking of mineralogical structure. The products are metal nitrates, elemental sulphur and nitric oxide.

Other minerals which have a significant effect on leaching are siderite ($FeCO_3$) and calcite ($CaCO_3$). These minerals react rapidly with nitric acid forming CO_2 gas which was responsible for the increased pressure in the autoclave observed during leaching. Fe and Ca-nitrate salts are water soluble, however, in later stage of oxidation, the generated sulphate ions cause the precipitation of $CaSO_4$. These reactions can be expressed by the following equations:

$$FeCO_3 + 3HNO_3 + 0.25\ O_2 \longrightarrow Fe(NO_3)_3 + 1.5\ H_2O + CO_2 \tag{7}$$
$$CaCO_3 + 2HNO_3 \longrightarrow Ca(NO_3)_2 + CO_2 + H_2O \tag{8}$$
$$Ca(NO_3)_2 + H_2SO_4 \longrightarrow CaSO_4 + 2HNO_3 \tag{9}$$

Other gaseous products were formed during leaching. Nitric oxide (NO) (equation 2 and 4) was formed in the absence of oxygen, but in the presence of oxygen was easily oxidized to nitrogen peroxide (NO_2). In water, NO_2 causes nitric acid to be regenerated. Thus, leaching carried out in an autoclave with oxygen served two purposes: (1) dissolution of metals at elevated temperature and (2) regeneration of nitric acid in situ.

The bulk of sulphide sulphur present in the concentrate was oxidized to elemental form (67%). The remaining sulphur was oxidized to sulphate form staying partly with the leach solution and partly precipitating as insoluble $CaSO_4$ and $PbSO_4$.

During leaching, the effect of the following parameters were studied:

    (1) Temperature
    (2) Nitric acid concentration
    (3) Various feed materials
    (4) Oxygen pressure
    (5) Pulp density

Initial tests (Table 4) showed that parameters (1), (2) and (3) were the most important. Results can be summarized as follows:

TABLE 4 Preliminary Study on Leaching of Jig Concentrate No. 1 and Flotation Concentrate No. 1 in Nitric Acid System; Analyses of Leach Residues and Extraction of Metals as Function of Temperature and Quantity of Nitric Acid in the System

(Extractions calculated on basis of leach residues and feed materials)

Conditions						Leach Residue Analysis (%)									Extractions (%)								
Test No.	Time min.	Temp (°C)	O_2 MPa	HNO_3 conc mL/100 g solids	Head conc.	Ag	Ni	Co	Cu	Bi	As	Fe	Pb	S	Ag	Ni	Co	Cu	Bi	As	Fe	Pb	S
1	120	125	1	33	Jig	6.87	0.37	0.29	0.35	5.77	4.96	8.98	2.28	4.31	28.0	83.7	82.0	48.6	36.2	35.7	13.2	5.9	27.1
2	120	125	1	50	Conc.	5.19	0.37	0.29	0.22	6.36	3.98	10.94	2.15	5.55	44.4	83.4	81.4	66.1	47.9	47.2	32.8	9.0	-
3	120	125	1	67	No. 1	0.86	0.05	0.04	0.01	4.64	n.a.	8.32	1.65	1.61	92.8	98.4	97.0	98.7	58.9	-	35.5	45.3	78.1
4	120	125	1	100		0.40	0.02	0.02	0.02	0.74	0.67	4.40	1.70	1.74	97.3	99.5	99.4	98.3	94.7	94.4	72.6	54.7	81.0
5	60	100	1	50	Flot.	2.60	0.08	0.08	1.77	0.99	4.12	17.4	6.13	16.0	0.1	91.7	92.4	73.1	0.1	2.1	1.9	0.1	14.6
6	60	100	1	75	Conc.	3.18	0.04	0.04	1.44	0.75	1.47	14.8	7.38	17.1	2.4	96.8	94.2	80.2	27.6	70.5	28.7	1.1	27.6
7	60	100	1	100	No. 1	3.59	0.03	0.04	1.11	0.63	1.05	13.3	8.60	11.8	10.2	97.9	97.1	88.5	55.3	84.1	51.8	1.9	57.3
8	60	100	1	125		3.33	0.06	0.07	1.73	0.83	1.30	12.8	9.70	12.0	29.3	95.6	95.4	82.6	41.8	80.9	54.9	2.7	57.7
9	60	50	1	125	Flot.	1.76	0.02	0.03	0.55	0.36	0.56	8.1	10.00	15.2	67.2	98.4	98.2	94.5	74.1	91.8	71.7	3.6	46.9
10	60	75	1	125	Conc. No. 1	2.60	0.03	0.05	1.27	0.39	0.73	11.0	9.52	16.3	46.9	97.8	97.3	87.8	74.8	89.7	62.8	4.6	44.9

Temperature. The effect of temperature was studied between 50°C and 125°C. On the basis of silver extraction, a higher temperature showed a beneficial effect on leaching using jig concentrate No. 2 and depressing effect on flotation concentrate No. 2 (Table 5). This strange behaviour of flotation concentrate was explained by analysing the solution for metals during leaching. Flotation concentrate No. 2 leached for 3 h at 75, 90, 100 and 125°C resulted in 50, 95, 72 and 43% Ag extraction respectively. Samples of solution taken during leaching at 100°C and 125°C were analysed for Ag, Ni, Co, Bi, As, Fe and S. The behaviour of metals presented in Fig. 1 showed that about 65% of silver and iron, and 80 to 90% of other metals were dissolved at zero time and the dissolution equilibrium was attained within 30 to 60 minutes. Zero time was defined as the end of nitric acid injection. Extended leaching at 100°C did not effect the metal concentration in solution, with the exception of a slight decrease of arsenic. However, at 125°C, the iron, arsenic, bismuth and silver concentration decreased significantly when leaching was extended beyond 30 minutes.

An x-ray diffraction analysis carried out on leach residue showed the presence of silver jarosite. The leach solution contained all components required for silver jarosite formation: Fe^{+3}, $SO_4^=$ and Ag^+. Jarosite precipitation takes place at a higher temperature and this explains the negative effect of temperature on the extraction of silver from certain types of silver concentrates, particularly those containing high iron and sulphur.

TABLE 5   Leaching of Various Feed Concentrates Alone or as Mixture in Nitric Acid System;
Effect of Temperature on the Extraction of Metals Calculated on Residue Basis

Conditions: 150 g/L solids; 1 to 3 h retention time; 1 MPa O_2 Pressure;
100 mL HNO_3/100 g solids; Charge 2 L

Feed Conc. No.	Variable Conditions		Leach Residue				Ag in soln. (g/L)	Extractions (%)		
	Time min	Temp °C	Wt g	Analysis (%)				Ag	Ni	Co
				Ag	Ni	Co				
Jig No.2	Head		300	8.16	1.74	1.64				
	120	50	163	5.70	0.05	0.03	7.8	62.1	98.5	99.0
	120	75	150	2.07	0.06	0.05	10.2	87.3	98.3	98.4
	60	100	146	1.11	0.05	0.02	11.0	93.4	98.7	99.4
	60	125	213	0.62	0.05	0.04	12.6	94.6	97.9	98.2
Flot. No.2	Head		300	2.70	0.88	1.15				
	120	75	140	1.38	0.01	0.01	3.2	49.7	99.6	99.7
	180	90	135	0.31	0.01	0.01	4.0	94.8	99.6	99.7
	90	100	160	1.41	0.02	0.02	4.2	72.2	98.9	99.1
	60	125	203	2.29	0.01	0.02	1.5	42.6	99.2	98.8
Jig No.2 Flot No.2 (1:2)	Head		300	4.52	1.17	1.31				
	180	100	136	0.32	0.02	0.01	6.8	96.8	99.1	99.8
	180	125	205	2.01	0.02	0.03	4.6	54.9	98.9	98.5
Jig No.3	Head		300	31.05	0.68	0.56				
	180	100	136	16.5	0.01	0.01	48.8	76.3	99.5	99.4
	180	125	132	3.2	0.01	0.01	54.0	95.5	99.5	99.5
Flot. No.3	Head		300	0.83	0.19	0.24				
	120	100	206	0.24	0.01	0.01	1.2	79.9	99.3	98.6
	120	125	201	0.06	0.01	0.01	1.2	95.2	98.3	98.6
Jig No.3 Flot No.3	Head		300	11.05	0.35	0.35				
	120	100	181	0.94	0.01	0.01	18.6	94.9	99.9	99.0
	120	125	185	0.65	0.01	0.01	18.8	96.4	99.1	99.0

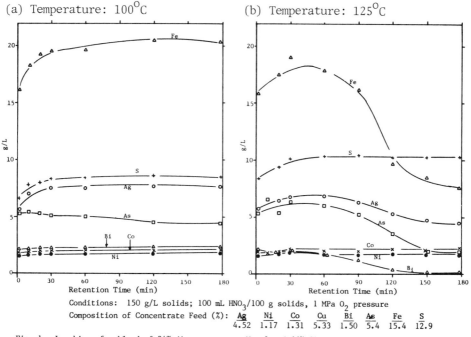

Fig. 1. Leaching of a blend of 34% jig concentrate No. 2 and 66% flotation concentrate No. 2 in nitric acid system; rate of metals dissolution at 100°C and 125°C

Nitric acid concentration. The requirement of nitric acid for leaching is dependant on the concentrate composition. Generally it can be stated that a sufficient quantity of nitric acid should be added to satisfy the formation of metal nitrates found in leach solution, with the same correction for sulphate in solution (~5 to 10 g/L S as $SO_4^=$). For simplicity, the quantity of nitric acid used in the tests was expressed as mL of concentrated HNO_3 (15.7 M) per 100 g concentrate. Nitric acid addition was studied in a range between 33 mL and 125 mL/100 g concentrate. Results showed that an increase of nitric acid from 33 mL to 100 mL increased silver extraction on jig concentrate No. 1 from 28% to 97.3%; and an increase of nitric acid from 50 mL to 125 mL increased silver extraction on flotation concentrate No. 1 from <0.1% to 29.3% as shown in Table 4.

The effect of nitric acid on metal extraction was also investigated on a blend of jig concentrate No. 2 and 3 and flotation concentrate No. 2 and 3 (4%, 5%, 24% and 57% respectively), at temperatures of 100 and 125°C. Conditions selected for these tests were: 30 min, 1 MPa oxygen, 150 g/L concentrate. Final residues were analysed for Ag, Ni, Co, Cu and Zn.

A graphic presentation of the effect of nitric acid on metal extraction is given in Fig. 2. The results showed that preferred conditions for silver extraction are 125°C and 125 mL HNO_3 per 100 g of concentrate. High nitric acid concentration apparently prevents formation of silver jarosite at 125°C. Extractions of Ni, Co and Cu remained high under all tested conditions, but zinc behaved similar to silver.

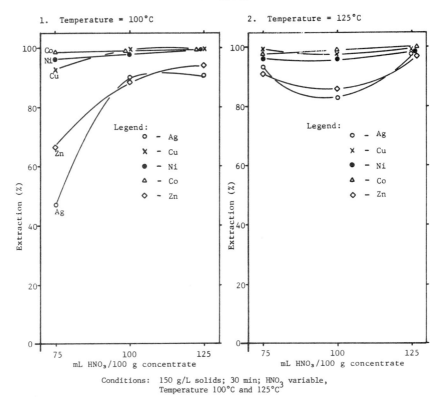

Conditions: 150 g/L solids; 30 min; HNO$_3$ variable,
Temperature 100°C and 125°C

Fig. 2. Nitric acid leach of blend of jig concentrate No. 2 and No. 3 and flotation
concentrate No. 2 and No. 3; effect of nitric acid concentrate on extraction
of metals at two temperatures: 100°C and 125°C

Various feed materials. The composition of silver concentrates differs widely and
the minimum quantity of nitric acid required for leaching has to be adjusted
accordingly. Leaching a blend of jig and flotation concentrates offers the best
alternative for treatment of this type of concentrate. Many tests carried out on
blend (Table 5) confirmed that high extraction of all metals is feasible.

Oxygen pressure. Oxidation of arsenides to arsenate and sulphides to elemental
sulphur or sulphate is carried out by nitric acid. Oxygen is required to oxidize
nitric oxide (NO) to nitrogen peroxide (NO$_2$). High overpressure of oxygen is not
necessary. The 1 MPa oxygen pressure used in tests was only to ensure the
availability of oxygen in the system. The vapour phase at the end of the leaching
tests contained 20% to 57% carbon dioxide.

Pulp density. Pulp density was investigated in a range from 150 to 500 g/L
solids. Solids in this range appeared to have an insignificant effect on the
extraction of metals, however, the filtration of leach slurry was much slower and
washing was more difficult. Therefore, most of the tests were carried out with
150 g/L solids.

Leach solutions. Typical leach solutions obtained in a series of tests with
various silver concentrates are shown in Table 6. Silver concentrations in these
solutions varied between 1 to 45 g/L. Leach solutions stored for a longer time
showed reprecipitation of some solids (1 to 2 g/L) comprising of Ca, As, Pb, Bi,
Si, Al and Fe. These solids were filtered off before precipitation of silver.

TABLE 6   Leaching at Various Feed Concentrates in Nitric Acid System;   Analyses of Leach Solutions
Prepared in Series of Tests on Various Feed Materials

Leaching Conditions: 50 to 125°C; 1 to 3 h retention time; ≃150 g/L solids,
1 MPa oxygen pressure; 50 to 125 mL conc HNO$_3$ per 100 g solids

Leach Soln. No.	Concentrate Feed	Analyses (g/L)													
		Ag	Ni	Co	Cu	Zn	Bi	As	Fe	Pb	Ca	Si	Al	S	U$_3$O$_8$
1	Jig Concentrate No. 1	12.5	4.15	2.88	1.05	n.a.	4.2	7.7	6.6	0.9	5.3	0.32	2.22	5.0	n.a.
2	Jig Concentrate No. 2	8.8	2.04	1.99	1.29	0.65	2.5	6.3	9.1	0.7	3.9	n.a.	n.a.	5.3	0.043
3	Jig Concentrate No. 3	45.0	1.00	0.83	1.12	–	1.0	1.9	7.4	0.2	n.a.	n.a.	n.a.	5.7	n.a.
4	Flotation Concentrate No.1	5.1	3.62	13.80	23.8	4.15	3.0	33.1	54.8	0.5	2.9	0.14	1.57	26.2	n.a.
5	Flotation Concentrate No.2	2.5	1.10	1.52	9.7	3.10	0.6	4.2	15.6	0.2	2.5	n.a.	n.a.	8.0	0.04
6	Flotation Concentrate No.3	1.0	0.22	0.29	2.9	n.a.	0.2	0.5	8.7	0.3	n.a.	n.a.	n.a.	4.1	n.a.
7	34% Jig Conc No.2 and 66% Flotation conc No. 2	4.7	1.36	1.15	6.8	2.38	1.1	4.0	11.4	0.3	4.8	n.a.	n.a.	7.3	0.041
8	34% Jig Conc No.3 and 66% Flotation conc No. 3	14.6	0.42	0.43	2.4	2.36	0.5	0.8	7.9	0.4	n.a.	n.a.	n.a.	2.1	n.a.
9	Blend of various conc.[1]	4.8	0.87	1.03	5.3	9.70	n.a	3.8	13.4	0.4	n.a.	n.a.	n.a.	4.1	n.a.

Note: 1) Concentrates: Jig Conc. No. 2   14%
Jig Conc. No. 3   5%
Flot Conc. No. 2   24%
Flot Conc. No. 3   57%

Leach Residue. Weight of the leach residue is about two-thirds that of the
original concentrate. It is comprised primarily of silica, alumina, lead
sulphate, calcium sulphate, elemental sulphur and iron. The minor components of
the leach residue are silver, arsenic, bismuth and zinc. Nickel, cobalt and
copper are in very low concentration. The valuable element present in leach
residue is silver (as unreacted silver mineral or as silver jarosite) and this has
to be recovered.

Leaching tests were carried out on a blend of leach residues obtained in the tests
listed in Table 5. This residue contained 1.77% silver.

The residue leached at 100°C, 125°C and 150°C for 30 min using 150 g/L solids, 150
mL HNO$_3$ per 100 g solids and 1 MPa oxygen, resulted in 87.4%, 94.0% and 94.7%
silver extraction respectively. This corresponded to 98.2%, 99.1% and 99.3%
overall extraction. The leach residue after second stage leach had the following
composition (%): 0.16 Ag, 0.025 Ni, 0.007 Co, 0.018 Cu, 0.068 Zn, 0.055 Bi, 0.30
As, 2.67 Fe, 10.2 Pb, 1.50 Ca, 0.91 Al, 21.9 Si, 11.5 S$_T$, 8.93 S°.

Recovery of Silver from Leach Solution

Many methods can be used for the precipitation of silver from the nitrate system,
such as, cementation, reduction with sodium borohydride, precipitation with
sulphide or chloride ions. Chloride ions were selected as the preferred method
because of selectivity, effectiveness and availability of reagents. Precipitation
takes place at room temperature and is instantaneous.

Tests were carried out on leach solution containing 40.8 g/L Ag and high
concentration of Ni, Co, Cu, Bi, As and Fe. The purpose of these tests were to
establish the quantity of chloride ions required for complete precipitation of
Ag. Sodium chloride was used as the precipitation reagent.

Results presented in Fig. 3 show that the silver was depleted from 40.8 to 0.0024
g/L by adding only a slight excess, over the stoichiometric quantity, of chlorine
ions. Increasing the chloride ion to 17.7 g/L increased the solubility of silver
to 0.07 g/L. Precipitation was very selective and only entrained impurities of

other metals were found in the AgCl precipitate. Careful washing of AgCl removed
these impurities. Using the chloride ions as reagent, silver was precipitated as
AgCl from all leach solutions listed in Table 6.

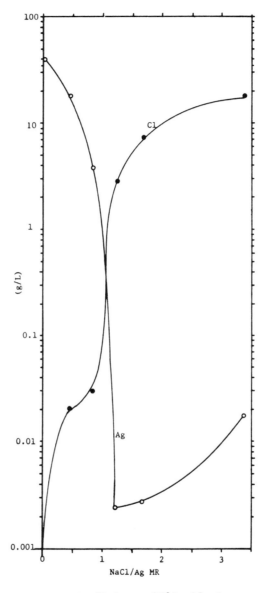

Conditions: 25°C, 15 min

Fig. 3. Precipitation of Ag from silver nitrate leach
solution with sodium chloride; Ag and Cl
concentration in filtrate vs quantity of reagent

Recovery of Bi, Fe, As, Cu, Ni, Co and Zn from Leach Solution

Silver free leach solution contains a considerable quantity of other metals, some
of which are valuable, such as Co, Ni, Cu, Bi, Zn and Pb and others without
commercial value, such as Fe and As. All these metals have to be recovered as
by-products or as safely disposable waste.

Exploratory tests carried out on silver free leach solution showed that by
neutralization of solution, various metals will precipitate at different pH level
(Fig. 4). Using the above technique, it was possible to separate the following
fractions: (1) bismuth precipitate, (2) arsenic-iron precipitate and (3)
Cu-Ni-Co-Zn precipitate.

Bismuth was precipitated at pH 0.4 to 0.8. In the presence of chloride ions it
precipitated as BiOCl and in the absence as Bi(OH)$_3$. Iron and arsenic
precipitated at pH 0.8 to 1.8. As the pH was increased above 1.8, copper, nickel,
cobalt and zinc also precipitated and then re-dissolved at higher ammonia concentration. It is preferable to quantitatively strip these metals as sulphides with
(NH$_4$)$_2$S. Technology for separation and recovery of Cu, Ni, Co and Zn from mixed
metal sulphide is well established[12],[13]. Lead was present in the leach
solution as a monor impurity (most of it remains with the leach residue) and it
was co-precipitated with arsenic-iron and bismuth fractions.

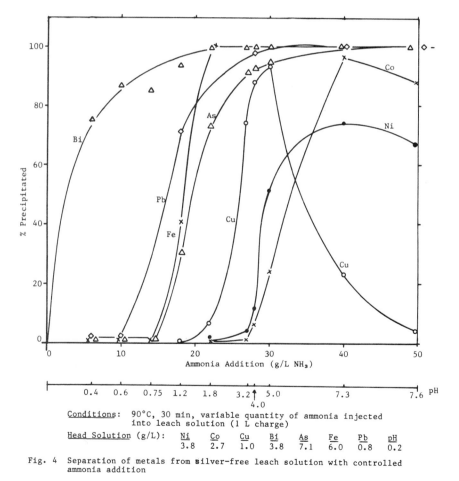

Fig. 4 Separation of metals from silver-free leach solution with controlled
ammonia addition

A large quantity of leach solution (62 L) was processed in ten batches at relaxed pH control. Analysis of products obtained in these tests are given Table 7. The bismuth fraction contained 41% Bi, the arsenic-iron fraction contained 20% As and 26% Fe, and mixed metal sulphides contained 15.9% Cu, 8.8% Co, 2.6% Ni and 3.1% Zn with <0.01% Bi, 1.1% Fe, 1.3% As and 0.006% Pb as impurities. The mixed metal sulphide residue, leached in sulphuric acid, resulted in solution containing (g/L): Cu = 47, Co = 26, Ni = 8, Zn = 9 and only 0.05 g/L As. Grades of precipitated fractions can be considerably improved by closer control of pH. Spent solution containing about 180 g NH_4NO_3, 50 g/L $(NH_4)_2SO_4$ and only traces of metals can be used for manufacturing of fertilizers.

TABLE 7 Analyses of Various Fractions Precipitated With Ammonia and Ammonium Sulphide

(Results from processing of 62 L solution in many batches)

Fraction	Method of Separation	pH	Analyses (%)								
			Ni	Co	Cu	Zn	Bi	Fe	As	Pb	S
	Head Solution		2.4	8.2	14.2	2.8	7.9	32.6	25.8	0.5	
Bismuth Precipitate	Ammonia	0.7–1.2	n.a.	n.a.	0.1	n.a.	40.7	8.0	8.2	n.a.	
Iron-Arsenic Precipitate	Ammonia	2.1–4.6	n.a.	n.a.	0.6	n.a.	2.9	26.3	20.0	n.a.	
Mixed Sulphide Precipitate	$(NH_4)_2S$	8.2–8.9	2.6	8.8	15.9	3.1	<0.01	1.1	1.3	0.006	24.4
Final Solution to be discarded (g/L)			0.001	0.001	0.001	0.001	0.015	0.001	0.026	0.002	10.9

Conversion of Silver Chloride to Metallic Form

Silver chloride is frequently an intermediate product of silver production. Usually zinc is used to convert AgCl to the metallic form, however, this product requires further purification. A recent publication[14] suggests that impure AgCl can be purified by dissolution in dimethylsufoxide and reprecipitated by dilution with water. Purified AgCl is converted to silver metal with zinc dust by reaction in aqueous suspension or by melting at 1100°C with an excess of Na_2CO_3. The other suggested method is direct reduction with hydrogen at 300°C to 400°C. However, application of all these methods in practice showed many disadvantages.

Tests showed that silver chloride can be purified from entrained impurities by washing with weak nitric acid solution (Table 8) and the purified AgCl can be converted to metallic silver by reaction with sodium carbonate at a temperature of 600°C.

A study was carried out on a mixture of AgCl and Na_2CO_3. This mixture was subjected to heat treatment in an air atmosphere. After heat treatment, the reaction product was washed with water and the silver metal and wash water were analysed for chloride. Studied parameters were: (1) temperature, (2) $Na_2CO_3/AgCl$ molar ratio and (3) retention time. Results given in Fig. 5 show that conversion of silver chloride to metallic form takes place between 500°C and 600°C within 1 h. About 1 mol of Na_2CO_3 is required to convert 1 mol of AgCl to metallic silver. The AgCl–Na_2CO_3 mixture reacted below 625°C was soft and easy to pulverize. Also, it did not react with porcelain dish. A temperature higher than 625°C caused sintering and produced hard cake. The following reaction took place during heat treatment:

$$2AgCl + Na_2CO_3 \longrightarrow 2Ag + 2NaCl + CO_2 + 0.5\ O_2 \qquad (10)$$

Products of the reaction are metallic silver powder, NaCl, CO_2 and O_2. The NaCl and an excess of Na_2CO_3 can be easily washed with water.

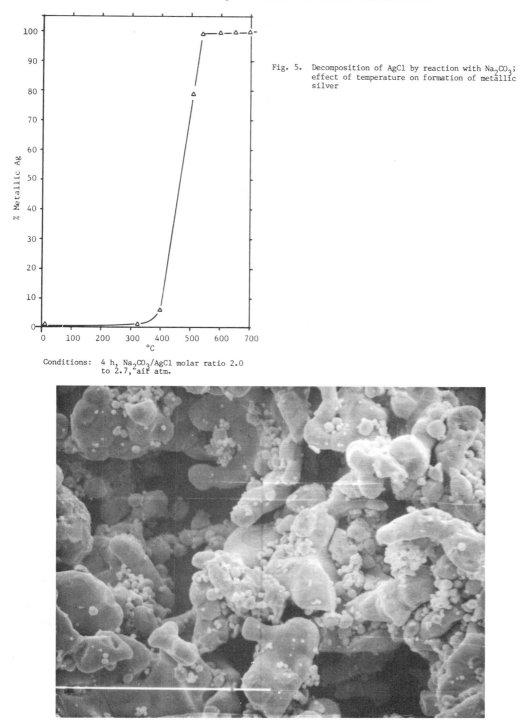

Fig. 5. Decomposition of AgCl by reaction with Na_2CO_3; effect of temperature on formation of metallic silver

Conditions: 4 h, Na_2CO_3/AgCl molar ratio 2.0 to 2.7, air atm.

x 2000 mag.

Fig. 6. Scanning electron micrograph of silver powder produced from AgCl salt by reaction with Na_2CO_3 at 600°C

W. Kunda

Chemical analyses and physical properties of a typical silver product are given in Table 8. The SEM micrograph in Figure 6 shows the shape and morphology of the silver product.

TABLE 8 Spectrographic Analyses of Water Washed and Nitric Acid Washed Silver Chloride and Silver Product; Physical Properties and Chemical Analyses of Silver Product

| | Spectrographic Analyses | | | Physical Properties and Chemical Analyses of Silver Product |
| | Silver Chloride | | Silver | |
Element	Water Washed	HNO₃ Washed	Product	
Ag	M	M	M	Physical Properties:
Al	0.04	X	X	A.D. = 2.5 g/cc
As	0.1	<0.01	<0.01	F.N. = 13.0
B	X	X	X	Surface Area = 0.8 m²/g
Ba	X	X	X	
Be	X	X	X	
Bi	0.08	X	X	
Ca	X	X	X	Chemical Analysis:
Cd	X	X	X	O₂ = <0.01
Co	0.03	X	X	C = 0.002
Cr	X	X	X	S = 0.001
Cu	0.008	0.001	0.001	Cl = 0.12
Fe	0.003	X	X	
Ge	X	X	X	
Hg	X	X	X	
Li	<0.01	<0.01	<0.01	
Mg	X	X	X	
Mn	X	X	X	
Mo	X	X	X	
Ni	M	0.002	0.002	
Pb	X	X	X	
Sb	X	X	X	
Si	0.03	0.003	0.001	
Sn	X	X	X	
Te	<0.005	<0.005	0.005	
Ti	X	X	X	
V	X	X	X	
Zn	<0.005	<0.005	<0.005	
Zr	X	X	X	

Flow Sheet

On the basis of the laboratory results, a flow sheet for the treatment of complex silver arsenide concentrate is proposed involving the following steps (see Fig. 7)

1. Two-stage countercurrent nitric acid leach.
2. Stripping of silver from leach solution with chloride ions.
3. Precipitation of bismuth with ammonia at pH 0.4 to 0.8.
4. Precipitation of arsenic and iron with ammonia at pH 0.8 to 1.8.
5. Precipitation of Cu-Ni-Co-Zn with sulphide ions at pH 5 to 7; the selective precipitation of Cu, Ni-Co and Zn fractions is also feasible.
6. Conversion of AgCl to metallic form by reaction with Na₂CO₃ at 600°C.

A two-stage countercurrent leach is suggested to obtain high extraction of silver and to utilize nitric acid more efficiently. Time required for leaching in each stage is only 30 minutes. Filtration of the first stage leach slurry can be replaced by thickener, with overflow solution to be processed and thickener underflow to be used as feed to the second stage leach.

Elemental sulphur can be extracted from second stage leach residue with (NH₄)₂S solution[15] and used for precipitation of Cu-NiCo-Zn sulphides. Lead, if present in large quantities, can also be recovered from leach residue.

The two-stage countercurrent leach was demonstrated in laboratory tests under conditions indicated in the flow sheet (Fig. 7). The results confirmed the expected extraction of all metals.

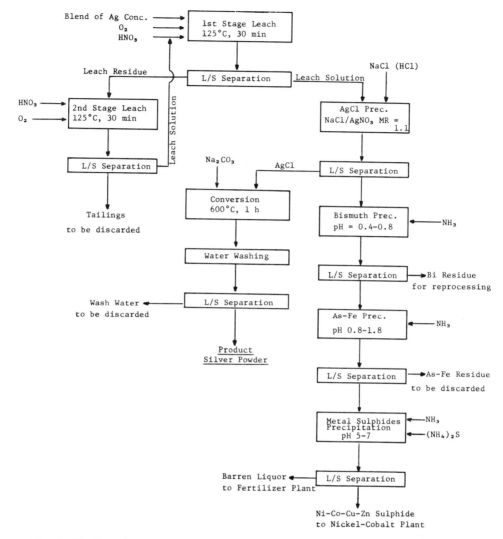

Fig. 7   Flowsheet for treatment of silver concentrate containing arsenic; from bismuth, nickel, cobalt, copper and zinc as major impurities

CONCLUSIONS

Silver concentrate, comprising native silver and bismuth and arsenides and/or sulphides of Ag, Bi, Ni, Co, Cu and Zn, responds to nitric acid leaching giving very high extraction of Ag, Ni, Co, Cu, Zn, Bi, As and Fe. The bulk of Pb present in the ore stays with leach residue and can be recovered as a by-product.

Under specific conditions, the dissolved Ag, Bi, As and Fe reprecipitate effecting the extraction of silver. This undesirable effect can be corrected by feeding the leach circuit with a blend of various concentrates to ensure more uniformity in composition, or by lowering the temperature or increasing HNO_3 addition. A two-stage leach will increase silver extraction to plus 99%.

Leach solution containing about 99% of all valuable metals present in the concentrate is further processed.

Silver is quantitatively precipitated as AgCl with sodium chloride or hydrochloric acid. Separation is very selective.

Silver chloride is converted to metallic silver by reaction with Na_2CO_3 at 600°C. This product is in the form of high purity silver powder.

Other metals are precipitated with ammonia and ammonium sulphide into three fractions: (1) bismuth precipitate, (2) arsenic-iron precipitate and (3) Cu-Ni-Co-Zn sulphide precipitate. Fractions (1) and (3) are valuable and can be further processed for recovery of Bi, Cu, Ni, Co and Zn. The final filtrate, containing ~180 g/L NH_4NO_3 and ~50 g/L $(NH_4)_2SO_4$ with only traces of other impurities, can be used for production of fertilizers.

The disposal of arsenic-iron residue should proceed in accordance with environmental regulations.

ACKNOWLEDGEMENT

This study was carried out at the Sherritt Research Centre in Fort Saskatchewan, Alberta. The author thanks the Management of Sherritt Gordon Mines Limited for permission to present this paper, and the Management of Terra Mining Exploration Limited for providing the silver concentrates.

REFERENCES

1. Silver Institute Letter, Vol. X., No. 1, January 1980

2. George, J.G., 1979. Silver, Can.Min.J., 100 (2), 84-91

3. Fridman, I.D., Shepeleva, K.A., and Berman, Ju. S. Effect of the Composition of Gold-Silver Ores on the Behavior of Silver During Cyaniding. Tsvetn.Met. 1976 (10), 70-2

4. Berezowsky, R.M.G.S. and Sefton, V.B., Recovery of Gold and Silver from Oxidation Leach Residue by Ammoniacal Thiosulphate Leaching. Paper presented at the 108th AIME Annual Meeting, Feb. 18-22, 1979, New Orleans, Louisiana

5. Panchenko, A.F., Kakovskii, I.A., Shamis, L.A. and Khmelnitskaya, O.D. Kinetics of Dissolution of Gold, Silver and Their Alloys in Aqueous Solutions of Thiourea. Izv.Akad.Nauk USSR, Met. 1975, (6), 32-7

6. Scheiner, B.J., Pool, D.L., Sjoberg, J.J. and Lindstrom, R.E., 1973. Extraction of Silver from Refractory Ores. U.S. Bur.Mines, Rep. Invest. 7736

7. Kunda, W., Hitesman, R. and Veltman, H., 1976. Treatment of Sulphidic

8. Kunda, W., Hydrometallurgical Process for Recovery of Silver from Silver Bearing Materials. Paper presented at the Third International Precious Metals Conference, May 8-10, 1979, Chicago

9. Ritcey, G.M., Ashbrook, A.W. and Lucas, B.H., 1975. Development of a Solvent Extraction Process for the Separation of Cobalt from Nickel. CIM Bull., 68 (753), 111-23

10. Queneau, P.B., and Prater, J.D. Nitric Acid Process for Recovering Metal Values from Sulphide Ore Materials Containing Iron Sulphides. Can. 995,468 (Cl 53-370; C.R. Cl), 24 Aug 1976, Appl. 163,413, 9 Feb 1973

11. Liddel, D.M., ed. 1945. Handbook of Nonferrous Metallurgy, 2nd ed., McGraw-Hill, New York, V.2, p 294

12. Kunda, W., Warner, P.J., and Mackiw, V.N., 1962. The Hydrometallurgical Production of Cobalt. Trans.Can.Inst.Mining Met. 65: 21-5

13. Kunda, W. and Mackiw, V.N., 1961. Sulphur and Sulphur Dioxide for Separating Copper from Nickel-Cobalt-Copper Sulphate Solutions. Can.J.Chem.Eng., 39: 260-4

14. Parker, A.J. and Clare, B.W. 1974. Salvation Ions. Same Applications IV. A Novel Process for the Recovery of Pure Silver from Impure Silver Chloride. Hydrometallurgy, 4: 233-245

15. Kunda, W., Rudyk, B. and Veltman, H. Recovery of Elemental Sulphur From Sulphur Bearing Materials. Paper presented at the Canadian Sulphur Symposium organized by the Inorganic Division and Calgary Section of the CIC, Calgary, May 30-31, 1974

Geology and Tonnage-Grade Relationships of Bulk Minable Precious Metal Deposits in North America Cordillera

Stanley W. Ivosevic
Consulting Geologist
P. O. Box 6568, Cherry Creek Station
Denver, Colorado 80206 U.S.A.

ABSTRACT

Bulk mining of large, low grade gold and silver deposits has been increasing in the North American Cordillera for about 20 years due to their ready exploitability.

Ideal deposits are thick subhorizontal bodies containing an average of 11 million T of ore.

Sixty-one deposits statistically analyzed are in five genetic classes. The classes and their average grades are: (1) Cenozoic volcanogenic, 3.85 oz/T eqAg (normalized to the average gold-silver crustal ratio); (2) Mesozoic volcanogenic, 2.4 oz/T eqAg; (3) subvolcanic, 0.066 oz/T Au; (4) Carlin-type, 0.140 oz/T Au; (5) sedimentary hosted, statistically misleading.

The average bulk minable gold deposit contains 8.5 million T of ore grading 0.110 oz/T eqAu; the average silver, 15 million T, 3.13 oz/T eqAg. Typical future developments are expected to be in 5-10 million-T gold deposits grading 0.09-0.175 oz/T eqAu and in 5-30 million-T silver deposits grading 2.2-5 oz/T eqAg; most developments are expected in Carlin-type and volcanogenic deposits.

Cumulatively, most bulk minable ore is in low grade deposits; but most contained metal, in relatively high grade ones. At optimum combinations of tonnage and contained metal, most gold ore and metal is in large, low grade deposits; most silver ore and metal, in medium grade deposits. There is little apparent tendency for size to increase toward some point of limitless reserves in more common genetic classes of deposits. Significant reserves can be expected in volcanogenic deposits, development of which class is just beginning to receive attention.

KEYWORDS

Gold; silver; bulk minable; geology; production; economics; grade-tonnage relations; ultimate reserves; North America.

INTRODUCTION

Bulk minable gold-silver deposits are those large tonnage, low grade precious

metals ore bodies enjoying circumstances of genesis and location which make them
commercially exploitable by large scale mining and metallurgical operations.

The term, "bulk minable (also called 'bulk tonnage') precious metals deposits" is
the writer's compromise between the cumbersome term, "large tonnage, low grade,
bulk minable precious metals deposits," and the frequently inaccurate "disseminated precious metals deposits."

The purpose of this address is to enhance the reader's perception of increasingly
important future Cordilleran sources of the gold and silver raw materials to be
utilized by refiners, dealers, and fabricators. The Cordillera is the chain of
mountains extending from Alaska, through western Canada and United States, south-
ward through Mexico to the southern tip of South America.

The geologic aspects of this topic will be discussed in detail only sufficient to
develop a rational framework within which to examine the grade-tonnage relationships which bear upon these future sources of supply.

Bulk mining of low grade precious metals ores is not a new development in the
Cordillera. For instance, by the middle 1960's, the Getchell mine, Nevada, had
produced over 1 million troy ounces of gold from bulk minable, low grade ore. Subsequently, there has been a general tendency for increasing numbers of bulk minable
gold or silver deposits to come on stream over the years.

Production from bulk minable gold-silver deposits particularly fulfills the goals
of large and small precious metals developers, because such deposits are readily
discovered, readily and economically exploitable, and more available than other
precious metals resources.

The average bulk minable gold-silver deposit contains about 11 million short tons
of ore grading up to 0.25 oz/T Au or 5 oz/T Ag. It is open pitted at a 4:1 stripping ratio at a rate around 5,000 tpd. It is either heap leached or milled; gold
is recovered at a high rate and silver, at a low rate. Capital costs range up to
$US 150 million (normalized to 1980 United Stated dollars).

GEOLOGY

Most bulk minable precious metals deposits identified to date have been of Carlin-
type gold or volcanogenic silver similar to the Delamar, Idaho, type. The former,
Carlin-type, are Paleozoic-carbonate-rock hosted disseminated deposits; and the
latter, volcanogenic types, are Mesozoic- or Cenozoic-volcanic-rock hosted epi-
thermal stockworks. Mesozoic volcanogenic deposits form in syn tectonic island
arc environments; and Cenozoic volcanogenic deposits develop in late tectonic,
extensional back arc environments. In places Carlin-type deposits and Cenozoic
volcanogenic deposits form consanguinously, in their respective host rocks, in the
same districts and from the same hydrothermal convection systems. Certain bulk
minable, intrusive-related precious metals deposits, herein designated subvolcanic,
originate in the foregoing environments also. Stockworks or stratiform orebodies
result which are sub horizontal, tabular, and thick.

Gold and silver not only occur as principle products of certain bulk minable
deposits but also occur in various proportions as co- or by-products with respect
to each other or with respect to another metal(s) or mineral product. Attributing
bulk minable qualities to by-product production does stretch the definition somewhat. Nonetheless, recognition is due under this definition to situations where
the sheer throughput volume of ores containing trace or low grade quantities of
precious metals place bulk minable operations for other commodities, such as the

copper operations at Bingham Canyon, Utah, and Butte, Montana, among the largest producers, respectively of gold and of silver in the United States.

GRADES AND GOLD-SILVER RATIOS

The following sections incorporate various aspects of the writer's statistical analysis of grade, tonnage, and operational data from 61 viable bulk minable gold and/or silver deposits of over 1 million T size in Canada, the conterminous United States, Mexico, Central America, and the Caribbean. This probably includes all of the most important bulk minable precious metals deposits (as opposed to unitized operations involving groups of smaller deposits) in the North American Cordillera at this time. Included are some deposits which were either recently depleted, temporarily abandoned, or are awaiting reactivation.

For convenience, the 61 deposits are distributed among the following five genetic groups:

      1. Cenozoic volcanogenic deposits. The average deposit grades around 0.06 oz/T Au and 2.7 oz/T Ag. Thirteen of the properties for which data are available contain silver as their major product and grade 3.85 oz/T eqAg. Equivalency is calculated by normalizing gold and silver to their average crustal ratio of 0.057 parts of gold to 1 part of silver

      2. Mesozoic volcanogenic deposits. These contain either major gold or major silver with local by-product base metals. Six of nine known examples are silver deposits grading an average of 2.65 oz/T eqAg; the remaining three are gold deposits grading a nearly equivalent 0.119 oz/T eqAu (which equals 2.08 oz/T eqAg)

      3. Subvolcanic deposits. All except one of 12 examples are gold deposits averaging 0.066 oz/T Au

      4. Carlin-type. Fourteen deposits grade 0.140 oz/T Au. These include, individually, some deposits which are unitized into larger operations considered in some of the following sections

      5. Sedimentary hosted. A statistically insignificant group, herein, of four concordant sedimentogenic silver deposits of limestone- and siltstone-sandstone affiliation

Cenozoic volcanogenic deposits contain higher individual and total grades of gold and silver than do Mesozoic volcanogenic deposits. In fact, gold tends to be negligible in Mesozoic deposits.

There appears to be a spectrum of deposits genetically transitional between Cenozoic volcanogenic and some subvolcanic deposits and continuing into Carlin-type deposits. The subvolcanic deposits generally are silver deficient.

Carlin-type gold deposits are characteristically silver deficient also. However, there are exceptions in which silver is present to contribute to the viability of Carlin-type mines.

The four silver-rich sedimentogenic deposits considered, although of low silver grade and gold deficient, occur with base metal co- and by-products. Although not representative of all sedimentogenic gold and/or silver deposits, this does serve to illustrate current trends in bulk minable precious metals development.

Most of the deposits group as being major gold or major silver with there being few

transitional instances of co-product gold-silver deposits. Thus, properties are
here classed as being major gold where their gold-silver ratio is greater than
0.057:1 (crustal ratio). By this method properties which industry generally
classifies as being gold deposits or silver deposits are clearly delineated.
Major product gold deposits are of the Carlin and subvolcanic types; major silver,
Cenozoic volcanogenic and sedimentary hosted types. Mesozoic volcanogenic deposits can be either major gold or major silver.

                    GRADE-TONNAGE RELATIONSHIPS

This section of the report successively addresses: (1) grade-tonnage relationships
among the various genetic classes of bulk minable gold, then silver, deposits; (2)
cumulative relationships; and (3) speculations concerning the ultimate resources
available from bulk minable gold-silver deposits.

Grade-Tonnage Relationships of Deposit Classes

In detail, the typical bulk minable gold deposit has a mean grade of 0.110 oz/T
eqAu; and a mean tonnage of 8.5 million T. In detail, these figures are representative of the 12 volcanogenic and 11 subvolcanic gold deposits considered herein.
As individual statistical groups and in the aggregate, the grade of Cenozoic volcanogenic, Mesozoic volcanogenic, and subvolcanic deposits declines with increasing tonnage from an initial grade around 0.15 oz/T eqAu, which is about half the
average grade of Carlin-type deposits, to become about constant at 0.075 oz/T eqAu.
However, average grades of Carlin-type deposits of various sizes vary somewhat
uniformly between 0.03-0.27 oz/T eqAu; this conclusion may result from widely
varying cutoff grades in different operations per metallurgical process employed.
Expectably, larger deposits are more rare than smaller ones.

The typical bulk minable silver deposit contains 15 million T of ore with a mean
grade of 3.13 oz/T eqAg. There is some tendency for grades to decrease with increasing tonnage from grades generally around 5 oz/T eqAg to around 2 oz/T eqAg.
This is deduced from data which includes many deposits, particularly two large
tonnage sedimentary hosted ones, whose economic viabilities are enhanced by non
precious co- and by-products. In criticism of my method, not portraying the effect
of these by-products upon equivalent grade causes this statement of grade to in-
completely represent economic circumstances.

A comparison of the grades of all bulk minable gold deposits and silver deposits
exhibits a mixture of populations representing the polygenecity of the input
variables. There is a crude bimodality, in part, resulting from the mutual exclu-
sion of gold and silver in any given deposit.

Cumulative Grade-Tonnage Relationships

Studies on cumulative frequency of grade and tonnage suggest that most gold ore remaining to be developed is in Carlin-type and volcanogenic deposits of 5-10 million
T size grading 0.175 oz/T eqAu and in Carlin-type, volcanogenic, and subvolcanic
deposits of comparable size and grading 0.09 oz/T eqAu. Generally speaking most
gold ore is in relatively lower grade deposits; and most contained equivalent gold,
in relatively higher grade deposits.

On the other hand, most silver ore occurs in the range from 2.2-3.8 oz/T eqAg; with
exceptions, this coincides with deposits in the 5 to 30 million T categories.

There are a substantial number of deposits of just under 5 oz/T eqAg of around 6 million T size.

Ultimate Resources

Various statistically and intuitively based predictions have been made on the con- tribution to ultimate global reserves of various metals from the development of deposits of increasingly larger tonnage but of lower grade to some point of un- limited reserves at which grade has diminished to approach the crustal average for the object metal. The statistical analysis herein contributes nothing definitive to this dialog, in part, because: (1) statistically significant populations were not sampled; (2) data from marginal deposits and those below 1 million T in size were omitted; (3) and the innate incomparability of grade-tonnage data.

However, the study permits these tentative conclusions having a bearing on ulti- mate resources. Most bulk minable ore is in relatively lower grade deposits, whereas most contained gold and/or silver is in relatively higher grade deposits. Although developmental economics favor larger deposits, smaller deposits of lower grade are more abundant than their higher grade larger counterparts.

A substantial amount of gold ore and gold metal is in larger tonnage and/or lower grade deposits. Most of this tonnage, is in volcanogenic and subvolcanic deposits. The postulated grade diminishment of the economically somewhat unimportant genetic group of subvolcanic deposits may be a reality. However, contrary to popular expectations, tonnage and grade of Carlin-type deposits (which are a major poten- tial resource) increase directly within the limits of the available data; tonnage of individual deposits probably will not increase significantly beyond that of known examples, but tonnage from individual operations on groups of nearby deposits will be in excess of what is already producing.

The ultimate limits of silver deposit development are less clear. Most silver ore is of medium grade, and grades do not appear to dimish toward some remote ultimate point of limitless reserves.

However, the writer believes that the geologic potential of bulk minable volcano- genic and related deposits is just becoming recognized and, particularly that the development of bulk minable volcanogenic gold and, to a lesser extent, volcanogenic silver deposits is only in its infancy. Therefore, the trends cited are subject to modification with the addition of data from future mine developments.

Recovery and Refining

Mechanical Processing of Electronic Scrap to Recover Precious-Metal-Bearing Concentrates

Fred Ambrose, Chemical Engineer, and
B. W. Dunning, Jr., Metallurgist

U.S. Department of the Interior, Bureau of Mines,
Avondale Research Center, Avondale, Maryland

ABSTRACT

A process research unit (PRU) for recovering precious-metal-bearing concentrates from shredded electronic scrap has been developed by the Bureau of Mines. The process operations include shredding, air classification, wire picking, magnetic separation, screening, eddy current separation, and high-tension separation. This semicontinuous system is capable of processing approximately 2 tons of scrap per day. The products include an aluminum fraction, a wire fraction, a copper fraction containing most of the precious metals, an air classifier light fraction, and a magnetic fraction.

Operation of the PRU is described with reference to obsolete military electronic scrap. Data on materials distributions and quality, potential recyclability, and further upgrading needs of the recovered products are described.

KEYWORDS

Precious metals; electronic scrap; mechanical processing; secondary recovery.

INTRODUCTION

As part of the Bureau of Mines mission to conserve our country's mineral resources through secondary recovery, researchers at the Bureau's Avondale Research Center are investigating mechanical processing techniques for recovering base and precious metals from electronic scrap. The objective is to develop methods of recovering these metal values from obsolete military electronic scrap currently not recovered in terms of adequate monetary return to the Federal Government. For example, during precious metals recovery operations from electronic scrap, aluminum metal may be lost completely, and low recovery of nickel, present in some copper alloys, results in a costly penalty. Upgrading of the electronic scrap to recover aluminum and nickel would result in a greater return to the Government in three ways. First, the aluminum and nickel recovered could be sold for their intrinsic scrap recycle value, and secondly, the quantity of material which would be processed for precious metals recovery would be reduced. Finally, this would result in lower toll charges for the same amount of precious metals recovered and would reduce penalties for the contained nickel.

Electronic scrap can be treated as a complex ore from which the various components
are to be concentrated in discrete fractions. Most of the precious metals are
associated with items such as pin connectors, contact points, silver-coated wire,
and terminals, whereas base metals, such as copper, iron, and aluminum, are primarily associated with components such as uncoated wire, transformers, and chassis. Based on laboratory-scale research conducted at the Avondale Research
Center, a process research unit (PRU) has been assembled that upgrades electronic
scrap into valuable metal concentrates at a feed rate of about 500 pounds per
hour. Detailed data on the recovery and grade of products will be published in
subsequent reports.

The Defense Property Disposal Service of the Department of Defense, under a memo-
randum of agreement with the Bureau of Mines, has supplied over 23,000 pounds of
scrap electronics which has been upgraded in the PRU. Evaluation of the metal
concentrates by private industry is in progress. Results of these evaluations
will determine future research goals by identifying those concentrates needing
additional upgrading. Initial results of the completed campaigns are encouraging,
creating the foundation for a new approach to the recovery of precious and/or
critical metals from electronic scrap.

PROCESS DESCRIPTION

The PRU is a series of unit operations designed to take advantage of the physical
properties of the various components found in electronic scrap to effect their
separation. The sequence of principal operations (Fig. 1) is shredding, air
classification, wire picking, magnetic separation, sizing, and, finally, eddy-
current and high-tension separation. The unit operations, with the exception of

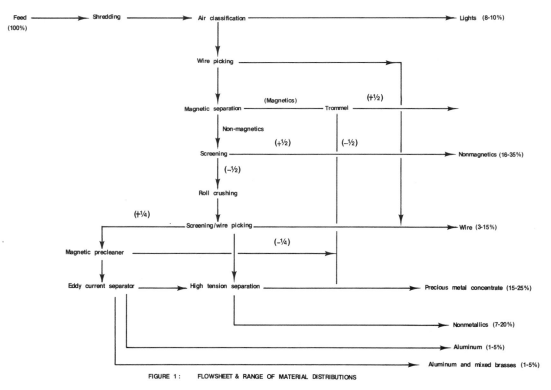

FIGURE 1:   FLOWSHEET & RANGE OF MATERIAL DISTRIBUTIONS

shredding, are not energy intensive and do not require significant labor input or specialized skills. All of the equipment, with the exception of the eddy-current magnetic precleaner, is currently in use throughout the mining and recycling industries and represents an "off-the-shelf" approach to the overall design of the PRU. This series of operations demonstrates the practicality of the method of metals recycling from electronic scrap.

The PRU operations have performed well on very heterogeneous scrap feed. However, visual observations of the scrap processed indicate that initial selective separation of incoming scrap by lots may prove beneficial. Care must be exercised to insure that scrap from unusual or special purpose electronics is not mixed and processed with general electronic scrap. A good example of this is electronic scrap where the entire chassis and nonfunctional components are plated with precious metals. During a trial campaign several small aluminum boxes, with their external surfaces painted flat black, were found after shredding to be triple plated, gold over silver over copper over the aluminum substrate. The gold content was approximately 1 percent of the case weight. Processing demonstrated that this material ended up in the aluminum fraction. Careful visual prescreening, i.e., the removal of the outer box cover, of the incoming electronic scrap eliminates this problem.

The following is a detailed description and discussion of each unit operation.

Shredding

The degree of success achieved in the actual mechanical separation processes is determined to a very large extent by the effectiveness of the shredding operation. A hammer mill,[1-2] having a 25-horsepower motor drive, 8-pound hammers, and equipped with steel punch plate grates having 1-1/4-inch holes on 1-5/8-inch staggered centers is used in the PRU. It is capable of shredding units having a volume of approximately one-half cubic foot at a feed rate of approximately 500 pounds per hour. Because of the size of the hammer mill, items larger than one-half cubic foot are dismantled and/or sawed to reduce them to an appropriate size.

During initial runs it was determined that processing massive items such as motor-generator sets and gyroscopes would result in the premature failure of the grates and subsequent excessive down time of the equipment. To eliminate this problem as the scrap is sized for feeding to the hammer mill, any unusually heavy items, or those identified as problem items, such as gyroscopes, are simply removed from the material processed. A sufficient number of gyroscopes, motor-generator sets, etc., were shredded to assess the liberation of the various materials used in their construction. Aside from catastrophic wear of the grate system, this operation was successful.

Air Classification

Air classification is carried out using a Bauer air classifier, Fig. 2, originally designed for use in the food processing industry. The unit was modified by replacing the air distributor screen with a solid, annular-shaped bottom plate to eliminate a plugging problem caused by wire entanglement and bridging over the discharge chute. Material discharged from the hammer mill is conveyed up to the intake chute of the Bauer unit. Here it falls onto a rotating distributor plate; lights are entrained in the airstream above the plate, and the heavies are channeled into a central discharge chute beneath the distributor plate. Lights are collected in a baghouse along with dust particles aspirated from the discharge chute of the hammer mill. The heavies are conveyed upward into a magnetic separator. A person standing next to the conveyor which moves the material to

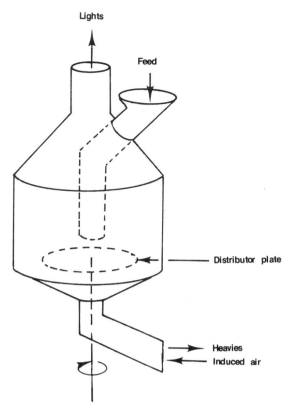

FIGURE 2 BAUER AIR CLASSIFIER

the magnetic separator picks off wire bundles that are generated inside the air classifier. Removal of the wire bundles eliminates bridging in the magnetic separator.

Magnetic Separation

An Eriez magnetic drum separator is used for separating the magnetic from non-magnetic components following air classification. These units are commercially available in a wide range of sizes and magnetic field intensities and product-handling capacities. Standard operating procedures were used.

Sizing, Magnetic Fraction

The magnetic fraction is conveyed to a trommel for sizing into minus and plus 1/2-inch fractions. The minus 1/2-inch fraction of the magnetic material contains such items as transistor caps and some magnetic connectors which may be plated

---

[1]Jeffery-Dresser Minimill.

[2]Reference to specific company or trade names does not imply endorsement by the Bureau of Mines.

with gold and/or silver.[3] Additionally, the transistor caps contain some 40 or more percent nickel, by weight.

Studies[4] at the Salt Lake City Research Center have demonstrated the feasibility of using this magnetic material to displace copper (cement copper) in solution. The gold- and silver-plated, nickel-alloy transistor caps are not affected during this operation and remain behind as an easily separated oversized material from the micron-sized cement copper. The Salt Lake City Research Center also smelted and electrolytically refined this material to recover the nickel, gold, and silver values. The oversize, plus 1/2-inch fraction is mainly a high-silicon steel which possibly can be returned to the steel industry for specialty steel use. However, tramp copper content may be too great to permit this, in which case this fraction would also be suitable for use in producing cement copper.

Sizing, Nonmagnetic Fraction

The nonmagnetic fraction exiting from the magnetic separator is conveyed to a vibrating screen separator where it is sized into plus and minus 1/2-inch fractions. The plus 1/2-inch material is primarily aluminum and plastics with some miscellaneous brasses, bronzes, and stainless steel. This material contains an undetermined quantity of precious-metal-plated parts, primarily as pin connectors still contained in molded plastic assemblies. Investigations are underway to determine a feasible approach to upgrading this fraction. The minus 1/2-inch fraction is conveyed to a roll crusher.

Roll Crushing

An Exolon roll crusher having 2-foot-diameter by 12-inch-wide manganese steel rolls is used on the minus 1/2-inch nonmagnetic fraction to flatten all the metal components and pulverize brittle plastic particles which still may be holding metal pins and connectors. The liberation of metal from metal-plastic composites and the increase in surface area of the small metal particles are the prime functions of the roll crusher. These two factors improve subsequent metal-nonmetal separation techniques. Material discharged from the roll crusher is conveyed onto a vibrating screen separator and wire remover.

Screening and Wire Removal

A modified two-deck Kason vibrating screen separator having an 18-inch-diameter working surface, shown schematically in Fig. 3, is used to separate the roll-crushed product. The upper screen has been replaced by a thin steel plate with a small discharge hole cut into it. A weir divides the working volumes into two separate sections of unequal surface areas. The discharge hole is located in the

---

[3]Analyses provided by the Bureau of Mines Salt Lake City Research Center indicate the minus 1/2-inch magnetic fraction from scrap obtained at the Davis Monthan AFB, contained 5.96 troy ounce/ton Au, 7.2 troy ounce/ton Ag, and 11.5 percent Ni, based on total sample weight of minus 1/2-inch fraction.

[4]Results of these studies have been summarized in a manuscript by H. B. Salisbury, L. J. Duchene, and J. H. Bilbrey, Jr., entitled "Recovery of Copper and Associated Precious Metals From Electronic Scrap," which is to be published as a Report of Investigations by the Bureau of Mines. The report will include a section on the copper cementation studies with magnetic electronic scrap.

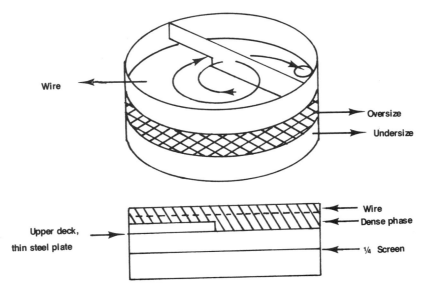

FIGURE 3 MODIFIED VIBRATING SCREEN

smaller area. An elongated opening on the bottom of the weir allows material to
pass from the large section to the small section. The material discharged from
the roll crusher is conveyed into the larger section. The weir prevents material
from exiting directly into the smaller section, causing it to circulate in a
tight circle. Fine particles and flattened metal migrate downward and form a
dense layer on which the wire "floats." The wire is periodically removed from
the large section by a worker assigned this operation; however, with additional
equipment modifications, this operation could be totally automated. The wire
from the earlier picking operation is combined with this wire. This wire is not
processed any further in the PRU. Accepted procedure for further upgrading this
type of wire is granulation and separation of the insulation from the metal with
an air table.

The dense material passing through the opening in the bottom of the weir drops
onto a 1/4-inch-square grid screen. This screen separates the flattened metals
from the crushed plastics and fine metal particles. The two screen fractions are
processed batchwise. The undersize (minus 1/4-inch) is processed through an
electrostatic separator, and the oversize is processed with the eddy-current
separation circuit.

Magnetic Precleaning

The plus 1/4-inch metal fraction from roll crushing contains some slightly magnetic work-hardened stainless steel that interferes with the operation of the
eddy current separator (ECS). Commercially available magnetic separators, such as
the Eriez unit discussed previously, will not adequately remove this weakly magnetic material.

Therefore, a magnetic separator using magnets similar to those used on the ECS
unit was constructed (Fig. 4). In operation, a 10-foot endless stainless steel
belt moves across an array of barium ferrite magnets. The belt and array are
inclined at a 45° incline. Material is discharged on the upper-left-hand corner

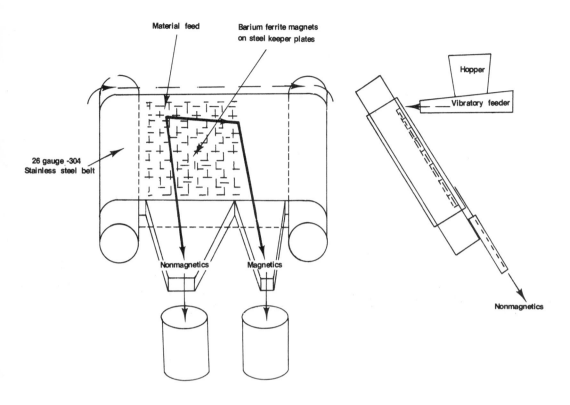

FIGURE 4   MAGNETIC PRECLEANER

of the belt, just below the top edge of the magnets. Any magnetic particles are
held fast against and move with the belt across and out of the magnetic field, and
fall down the inclined belt into a collection funnel. The nonmagnetics are un-
affected by the magnets and slide directly down the slowly moving belt into a
second collection funnel. Materials discharging from the funnels are collected in
drums to await further treatment.

The magnetic fraction thus collected contained a significant number of plated components and is a concentrate for precious metal recovery. Nonmagnetics are processed on the eddy current separator.

Eddy Current Separator

The eddy current separator (ECS), developed and patented by the Raytheon Corp.,
U.S. Patent 4,003,830, has the appearance of a large, inclined draftsman's table.
A 10-foot by 10-foot stainless steel (nonmagnetic) skin covers a bed of magnets,
which are arranged in rows of alternating polarity. Each magnet has its magnetic
field oriented along the axis normal to the plane of the table. The table surface
is covered with rows of diagonally positioned magnets. A steel (magnetic) plate
under the magnets acts as a "keeper plate" to hold the magnets in the proper orientation. The face of each magnet measures 1/2 inch x 1 inch and is 3/8 inch
thick.

In practice the operation is quite simple and is based on the principle that a conductor moving through an external magnetic field will have an electric and magnetic field induced within. The action is analogous to what happens in an induction motor. The induced magnetic field and the primary magnetic field repel each other, causing a resultant force on the conductor and the magnets. Since the mass of the conductor, the flattened metal particles, is many orders of magnitude smaller than the magnetic array, the entire resultant force can be assumed acting on its mass alone. The magnitude of the resultant force is a function of many parameters, e.g., particle velocity, electrical conductivity, mass density, and magnetic properties of the particle.

The size of the magnet's pole face will determine the smallest particle that can be separated. In practical terms the smallest particle size that will be efficiently deflected is approximately the same size as the pole face, although some deflection is noted down to where the particle size is approximately half the width of the pole face.

Very good separation of aluminum from miscellaneous brasses and bronzes and miscellaneous brasses and bronzes from stainless steel is possible. While pure copper and aluminum particles of the same diameter and thickness experience essentially the same deflection on the ECS, there are few components in electronic scrap (with the exception of wire) that are fabricated from pure copper. Therefore, the aluminum fraction is quite clean.

ECS separation has proven very successful on the plus 1/4-inch, minus 1/2-inch material which has been roll crushed prior to the second screening. There are three clearly definable fractions that report from this process. Materials deflected the farthest along the board are aluminum alloys and pure copper. Aluminum, representing better than 26 percent of the total feed material, deflected from 4 to 8 feet. Materials deflecting from 2 to 4 feet are represented by nickel silvers, brasses, and bronzes. This material amounted to 18 percent of the feed. The remaining material feed experienced little if any deflection (0 to 2 feet) and was composed of nonmagnetic stainless steel, nonmetals, small metal particles, and nonmetal-metal composites. The 0- to 2-foot fraction contains many precious metal-plated parts, primarily connector ends and pins. Further upgrading of this material is carried out in high-tension separation (HTS), where nonmetals are removed.

The nonmagnetic, plus 1/2-inch oversize may be upgraded on the ECS as well; research into this processing option is currently underway. Connector assemblies, pin connector terminals, etc., from which the plated parts have not been liberated, will not experience any deflection on the ECS, while the large aluminum, brass, and bronze pieces should experience deflection. This would reduce the material that must be processed for precious metal recovery and increases the quantity of clean base metal that can be sold for its intrinsic value.

High-Tension Separation

High-tension separation is a method based on electrostatic attraction/repulsion and the ability of a nonconductor to hold a charge and a conductor to lose a charge when in contact with an electrically charged surface in an external electric field. A clean separation of metallics from nonmetallics is attainable using a Carpco high-tension separator in this operation.

The electric field is imposed across a pair of electrodes and a rotating drum; the drum is positively charged and the electrodes negatively charged. The material is conveyed onto the rotating drum and attains a radial velocity approaching

that of the drum. The electric field charges all material negatively. Since
unlike charges attract, the negatively charged nonconducting material is attracted
and held to the rotating drum. As the drum rotates, conducting material loses its
negative charge quickly, assuming the same charge as the drum, resulting in an
electrostatic repulsive force between the particle and the drum. The combined
electrostatic repulsion and radial velocity cause the particle to assume a bal-
listic trajectory which is determined by the rotational speed of the drum and the
field strength. A knife edge splitter directs the metallic fraction to a discharge chute. The metallic fraction will contain most of the silver- and gold-
plated components. A portion of the nonmetallics rotates past the knife edge and
falls free of the drum. This is primarily due to the shape and weight of those
particles. This is a middling fraction that contains some metal. The remaining
nonconducting material is primarily finely divided thermoplastics and thermo-
setting plastics which must have their charge removed in order to drop free of the
drum. This nonconducting fraction will contain a high percentage of copper in the
form of extremely fine, short pieces of hair wire that are generated during the
shredding procedure. Nonmetallic pieces mask and hold the fine wire to the drum,
negating the repulsive forces acting on the wire. Initial visual indications were
that there was relatively little precious metal in the nonmetallic fraction.
Assay of this fraction proved that visual observation of this fraction is not
reliable. It is possible to reduce the total visable metallics to less than 0.1
percent in the reject material through repeated passes through the HTS, and fine
tuning the HTS (e.g., using different settings for electrode position, field
strength, rotor speed, etc.) for each pass. Also, different electrode configu-
rations may result in a better separation. However, an in-depth study of this one
operation is beyond the scope of this research.

The precious metal content of the metal and nonmetal fractions from HTS are shown
in Table 1 for typical, state-of-the-art electronic scrap. The feed input to HTS
was found to contain 24.67 troy ounce/ton of gold and 296.41 troy ounce/ton of
silver.

TABLE 1   Precious Metal Content of HTS Fractions

Fraction	Au, troy ounce/ton	Ag, troy ounce/ton
Metal	42.37	505.80
Nonmetal	3.63	47.41

The material not significantly deflected on the ECS (0-2 feet) is further upgraded using HTS. Characterization of the nonmetal after two passes showed less
than 0.25 percent of the plated parts present.

Reducing the plastics present in the material bearing the precious metal lowers
the emissions problem from the furnaces burning this material, prior to melting.
However, there is a need for a special facility with a complete emission controls
system to handle the lower grade nonmetal fraction. With the large surface area
of the high-tension separator products, hydrometallurgy may be the best and
cleanest approach to a new refining process, eliminating the need for burning and
melting operations.

CONCLUSIONS

General electronic scrap can be treated in a manner analogous to an ore: bene-
ficiated, upgraded, refined, and returned to the manufacturing circuit. Using
existing technology, adapted to the characteristics of a particular material, unit
operations can be integrated into an automated flowsheet. Electronic scrap, where
the precious metals are confined to plating on connector assemblies, printed circuit cards, wiring, and the like, benefit greatly from this procedure by reducing
the weight of material to be processed. The current method of hand stripping of
electronics is slow, unpredictable, expensive, and subject to wide variation in
performance. An automated process as described in this report is predictable,
not labor intensive, and uses technology solidly established in many other
industries.

In the general case, where everything from relays to waveguides is processed
simultaneously, then the operations described in the text can be expected to
result in a minimum amount of material that must be processed for precious metals
recovery. The fractions generated are an iron-base fraction, an aluminum fraction, a mixed metal fraction, a high and a low precious metals concentrate, a
light fraction, and a wire fraction.

The plus 1/2-inch iron-base fraction will contain some copper, and the minus 1/2-
inch fraction may also contain some nickel and precious metals. If the entire
iron-base fraction were used to cement copper, the copper contaminant would be
additive, and the nickel and precious metals could be recovered from the residue.

The aluminum fraction of the ECS represents a high-value base metal product
directly recyclable into the secondary aluminum market. It could be further upgraded into copper-free aluminum scrap by a heavy-media treatment. The mixed-
metal fraction might also be upgraded further using conventional heavy-media
techniques. Testing is continuing on this material to determine processing techniques for maximum recovery of the metal values.

The wire fraction should be directly processed for copper and precious metals
recovery. There are several options for this which include chopping and further
upgrading, melting, casting and electrorefining, or hydrometallurgy.

The "lights" from air classification and dust control would represent a potential
"lower value" copper and precious metals concentrate.

The precious metals concentrates from HTS and magnetic precleaning may be treated
either by conventional pyrometallurgical and electrometallurgical techniques or
by newly developed hydrometallurgical techniques.

A Printed Wiring Board Manufacturer Looks at Precious Metal Utilization and Recovery

A.W. Castillero

Xerox Corporation, Pomona, California

ABSTRACT

Precious metals have been used extensively in the manufacture of printed wiring boards. In general, the economic conditions of the past few years have dictated the use of the thinner and more selective precious metal deposits, and the adoption of measures to curtail waste. This paper reviews metods of prevention of losses due to overplating and dragout, and discusses the reclamation of scrap and reject materials.

KEYWORDS

Printed wiring boards; contacts; electroplating; waste; dragout; recovery.

INTRODUCTION

Precious metals have been used extensively in the manufacture of printed wiring boards. Many years ago it was common, for instance, to use connector fingers on printed wiring boards with nickel and rhodium plating. Rhodium plating usually ranged in thickness from 20 to 50 millionths. Because of the hardness and brittleness of rhodium, more and more printed circuit designers started to specify gold plating on the fingers for electrical contact requirements.

In the early days of the use of gold, it was common to call out for thicknesses of 200 millionths of an inch over the copper traces. When nickel was used as a barrier layer so that copper would not migrate through the gold, thicknesses of gold began to drop to 100 millionths, then in the last few years to 50 millionths on the fingers. The present Mil Spec has this requirement. Also, several years ago it was very common to plate 30 to 50 millionths of gold all over printed wiring boards. However, when gold started to escalate in price in 1972-1973, most printed wiring manufacturers started to turn to other metals as a final finish for printed wiring board conductors. One of the very common finishes that is presently used is solder plate, a tin/lead alloy of 60% tin and 40% lead. However, because of the contact resistance problems that would be encountered with tin/lead plating with the majority of present connectors in use today, nickel/gold is still used on the fingers of printed wiring boards to a large extent. Many manufacturers, however, are looking at going to much thinner coatings. Paradoxically, because of the rise in the cost of labor, some manufacturers have gone back to the use of gold all over but using only a flash with very low thicknesses. There are several companies, including ourselves, that are putting on a gold flash of from three to eight millionths all over the board which serves as the final finish for the board. In this particular instance, I might mention that with present day etchants, the gold plating does not act as the etch resist in the low thickness, but rather we rely on the nickel underneath to protect the copper from attack in an alkaline etchant system. This process, at least until a few months ago, was less expensive than the process of solder plating and its accompanying intensive labor operations required (1) to mask above the finger; (2) chemically strip off the solder from the tips; (3) remask, in some instances; (4) nickel/gold plate the tips; and (5) remove tape and clean.

Now, with a still much higher price of gold, printed wiring board manufacturers and designers are looking at other alternatives. Many designs now eliminate the use of edge connectors and go to a two-piece connector instead. Also, many manufacturers are investigating the use of other precious metals that have a lower density, therefore, such metals as palladium and palladium/nickel alloys are receiving considerable attention. Equivalent thicknesses of palladium would cost one fourth the cost of gold.

I do not, however, anticipate that these new alternatives will change a requirement for gold plating overnight.

Therefore, we are faced with a continuous problem of effective use of the precious metal gold, its economical application, and its recovery. The first thing that is essential in the electoplating of gold on finger connectors for printed circuit boards is that the material not be wasted. In other words, you should not apply more gold on the etched connectors than the specification minimum requirements. Some gold is wasted by inadequate controls, improper anode configuration, and a lack of knowledge of the plating process.

First of all, it is essential that the printed wiring shop be equipped with a testing device so that they can measure the thickness of gold they are applying. A common method of testing the gold thickness is to use a Beta-backscatter apparatus of which there are two or three on the market. I am sure most of you are familiar with this equipment, but amazingly enough, many paople do not use it.

If you are finger plating tips only, it is important that the anode configuration you decide for your tank be such that it gives you good distribution. If your requirements are for 50 millionths minimum, it is common to shoot for 60 millionths, but anything higher than that would be a waste.

It is not uncommon, in some instances, on the edges of a board to reach thicknesses twice that of the requirement.

To address that problem, there are machines now on the market that mask the printed wiring board by the use of conveyor belts in a continuous line, going through the stripping, the nickel plating, and the gold plating on tips, the steps outlined before, achieving excellent results as far as distribution is concerned. It is reported that distribution ranges within five millionths of an inch are common. For large manufacturers of printed wiring boards the pay back of equipment of this nature is soon realized.

The second area where one must look to save gold is in the proper rinsing and utilization of dragout tanks. It is not uncommon in many small shops and large shops as well, to see no dragout tanks following the gold plating tank. It is estimated that perhaps 10 to 15% of the gold that is used in an operation is dragged out and rinsed away, therefore, it is very essential to have proper dragout tanks after the plating operation. It is usually a good idea to have at least two dragout tanks and to have some gold recovery device attached to these tanks.

There are several methods of recovering the gold from solution in these dragout tanks. One is the use of a resin bed that circulates the low concentration solution through resin which absorbs the material up to a certain point and then the resin can be exchanged, burned, and the gold recovered.

A second method is to electroplate the metal out of the solution. This usually requires the addition of some electrolyte.

A third method that is being proposed for use is electro-dialysis. This method is being used predominantly for nickel and chromium but now it is also being proposed for gold recovery.

A fourth method and one that we are working with is the use of an evaporator. The dragout solution is pumped into an evaporator column. Under low heat and vacuum, the water is distilled and is returned to the dragout tank. Thus, this provides one of the most economical methods of recovering the gold from the dragout solutions. We are, however, experiencing some sludge formation which may be the result of the breakdown on some of the bath ingredients. Also, the build up of contaminants may occur faster since you are returning the concentrate back to the bath.

Next, we shall look at the recovery of gold from the subsequent fabrication steps. Whenever tips are plated on a printed wiring board, these tips must be connectd by a tie bar at the bottom of the board and, in some cases, also have surrounding borders. This is material that does not end up on the finished board and therefore is scrapped. This material can be handled in a couple of ways. If we have large borders and end up with a frame, the frames and the tie bars can be stripped in a chemical solution, and subsequently recovered. Putting the gold in solution allows the user and the refiner to analyze the solution and come to some agreement on the amount of gold contained. If the tie bars are such that handling during the stripping operation would be awkward, then the best method is to burn the scrap to recover the gold. This, of course, leaves no easy way for determining the amount of gold present on the scrap material. This situation exists also in the case of rejected whole boards that are determined to be scrap.

If the board does not have any solder mask, it can be easily run through a stripping solution to remove the gold. This can be done in a barrel operation or it can also be done in a spray conveyorized machine, such as an etching machine. If the board, however, already has solder mask, it is very difficult to strip that portion of the gold underneath the solder mask by chemical means. In such cases, it must invariably be burned.

This also becomes the case in assembled boards that must be crushed and burned. In this case, the gold contained on some components is thus also recovered. You can, however, shear off the tips and strip those in a chemical stripper. In conclusion, there are effective ways of controlling the application of electroplate on printed wiring boards, and secondly, there are effective ways of recovering that gold that is applied on unwanted areas or that becomes part of the scrap of the manufacturing process.

As gold becomes more and more expensive, undoubtedly more people will begin paying closer attention to this costly operation.

Secondary Refining of Gold and Silver — The New Johnson Matthey Chemicals Facility

F.T. Embleton

Johnson Matthey Chemicals Ltd., Brimsdown, Enfield, Middlesex, England

ABSTRACT

This paper relates to the new facility of Johnson Matthey Chemicals at Enfield, U.K. and deals with the design requirements for a modern precious metals recovery plant.

Fundamentals such as security of values, accuracy of sampling, commercial integrity, control of the environment, compliance with the Safety at Work Act and high recovery rates are dealt with.

A description of the plant used for the preparation and evaluation of the wide variety of primary and secondary precious metal containing materials is followed by the processes used for smelting, pyrometallurgical and electrolytic refining.

INTRODUCTION

In the design of the new facilities at Brimsdown for Johnson Matthey Chemicals Limited, prominence was given to the following fundamentals:-

a) Security of client's materials after receipt along with accurate weighing.

b) Care of values in the preparation stages.

c) Accurate sampling and analysis.

d) Complete integrity in the commercial exchange of analyses.

e) Improved working conditions.

f) Safety and control of the environment, both within and around the factory.

g) High recovery of values during the refining processes.

The new plant, combined with the company's Royston refinery, provides the world's largest and most technically modern precious metals refining facility to treat the wide spectrum of available primary and secondary materials, including:-

Gold and silver bullion, dore	Liquid gold residues
Demonetised coinage	Electronic scrap
Jewellers sweepings	Photographic emulsions, film and paper
Lemels	Plating solutions
Carat gold scrap, silver scrap	Spent industrial catalyst
Rolled gold	Copper tankhouse slimes

Fig. 1   General Exterior View

PROCEDURES AND PLANT

Materials Control

After initial weighing on a 60 tonne Avery loadcell weighbridge, sweeps and low grade scrap are check-weighed against advices on a 5 tonne Toledo electronic scale which has the facility for the direct entry of weights and details into a computer.

Bullion and high grade scrap are weighed either on a 40,000 ounce troy Avery printout scale or on a 2,000 ounce bullion balance.

Details and weights are transmitted by land-line to the company's commercial department at Royston where customer references and requirements are added.

Working papers and stock cards showing all details are then printed ready to accompany the material through the various evaluation processes. Customers' materials are held either in safes or secure storage areas until required for evaluation.

Evaluation

With the present high intrinsic values of gold and silver, accuracy of evaluation has reached a

Fig. 2   Scrap Coin

Fig. 3   Electronic Scrap

new level of importance. Slight inaccuracies can cause losses which are sufficient to outweigh any profit margin made from the lengthy process of refining.

Sampling

Two basic routes are followed. Metallic materials are melted, whereas materials such as jewellers sweepings, photographic residues and tankhouse slimes are subjected to numerous processes in order to render them suitable for blending and sampling.

Metallic Materials (e.g. bullion and coin)

In order to obtain an accurate sample it is first necessary to melt the material, the sample being taken from the molten metal.

The Melting House is equipped with Inductotherm induction crucible furnaces with capacities ranging from 4 kg. - 500 kg. at powers of 30 kW to 200 kW.

Either silicon carbide or clay graphite crucibles are used to obtain "hot or cold pot" conditions according to the melt requirements.

All furnaces and casting stations are equipped with fume extraction hoods which are connected to a Tilghman bag filter of 30,000 cfm capacity. The vertically hung terylene filter bags are mechanically shaken and the collected dust is discharged into sealed containers ready for recycling. The filtered gases are discharged to atmosphere via a 100 ft. stack.

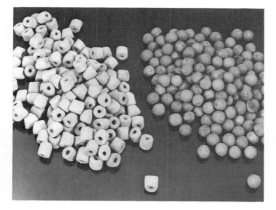

Fig. 4   Catalyst Scrap

Fig. 5   Melting House - General View

The decision to use induction furnaces was taken with the following points in mind:-

a)   Speed of melting.

b)   Higher attainable temperatures.

c)   Absence of products of combustion giving a lower required exhaust volume.

d)   The vigorous stirring action caused by the electromagnetic field assists in obtaining a homogenous melt from which the sample can be taken.

The melt is sampled either by dipping with a graphite spoon, from which buttons or granules can be cast, or by use of a glass vacuum tube to produce a thin rod of metal.

After casting into either anodes or bars, the metal is weighed and stored until released for processing.

Fig. 6   Melting House - Large Furnace Pouring   Fig. 7   Universal Incinerator SF3

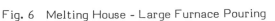

Sweeps and Residues including Tankhouse Slimes

In order to obtain accurate samples of these generally heterogeneous materials, it is necessary to carry out one or more of the following processes:-

Burning:

To complement the existing incinerators used for burning waste photographic film and paper, three new incinerators have been installed:

a)   Beverley A39 Comptro

b)   Universal SF3

c)   Universal SF5

All are gas fired and fitted with afterburners. Burning can be carried out either on the hearth of the furnace or, in the case of small lots, by loading in a bank of mild steel trays.

Drying:

A Fielder stream jacketed pan dryer has been installed for drying very wet sludge and press cakes. Low moisture materials are dried by means of a Calmic fluid bed dryer and an electrically heated tray dryer.

Grinding:

After burning or drying the resultant ash is ground to 30# in one of four Edgar Allen "Stag" mills fitted with integral screens through which the powder is discharged continuously into drums. Any metallics present remain inside the mill and are removed at the completion of the customer lot. These metallics are melted and treated separately from the bulk.

Very small, high value or difficult lots, are ground in a small Loftworth batch mill or a Bolton pan mill.

All mills are fitted with sound-proof enclosures, reducing noise levels in the vicinity to less than 85 dB(A).

Blending:

For blending the powders prior to mechanical sampling, two 1 tonne Vrieco Orbital screw blenders with mechanical charging systems have been installed. These discharge into cubic containers fitted with "Bomb door" type openings and having a capacity of approximately 1 tonne of sweep.

Fig. 8   Vrieco Blender

Mechanical Sampling:

The 1 tonne container of sweep (powder) is lifted to the top of a custom built "twin stream Knight sampling train".

The discharge doors are opened pneumatically and the powder flows into a vibratory hopper, giving a constant flow of material through the first Knight sampler which takes two separate 10% samples. Each of these 10% samples is fed to another Knight sampler which takes a further 10% cut.

The two resultant 1% samples of the bulk, each weighing $\pm$ 10 kg, are passed to the final sampling room for further reduction.

Coning and Quartering:

Certain powders, including those with bad flow characteristics, are still sampled by the classic coning and quartering method.

A feature of the grinding, blending and sampling area is that each piece of equipment is fitted with its own dust control unit. This ensures that any collected dust can be restored to the customer lot concerned.

Final Sampling:

The equipment installed in the final sampling rooms ensures that the sample passed to the laboratory is representative of the bulk from which it came.

Moisture is determined on one sample in the usual way.

The final sample for analysis is reduced to 2 kg using an SR40 rotary riffle. After grinding to -125# in a Tema vibratory mill, any +125# metallics left are melted for analysis. The powder is further reduced in volume on an SR5 rotary riffle in order to give sufficient samples of 100 grammes each for analysis. This riffle may also be used to give the assayer a catch-weight sample for the actual analysis.

Silver on Alumina Catalyst:

Spent ethylene oxide catalyst is sampled on a vacuum-fed mechanical sampling train.

The mixture is fed onto a double deck vibrating screen which separates the three fractions: support, fines and catalyst. The barren support is held in reserve and the fines are sampled by coning and quartering. The catalyst pellets, 8-14% silver, are mechanically sampled to give two separate 5% samples. One 5% is ground and sampled for analysis whilst the other is held in reserve.

Analysis

The laboratory is equipped to deal with the analysis of sweeps and bullion as well as control analyses. In addition to "paid for" metals, the samples are also assayed for metals carrying penalties or, in the case of sweeps, the make up elements such as silica, iron oxide, lime and alumina which are required when calculating flux additions for the blast furnace charge.

A wide variety of techniques can be used, ranging from a fire assay and atomic absorption to x-ray fluorescence and inductively coupled plasma. The latter is a direct reading emission spectrograph with a radio frequency inductively coupled argon plasma source unit. This

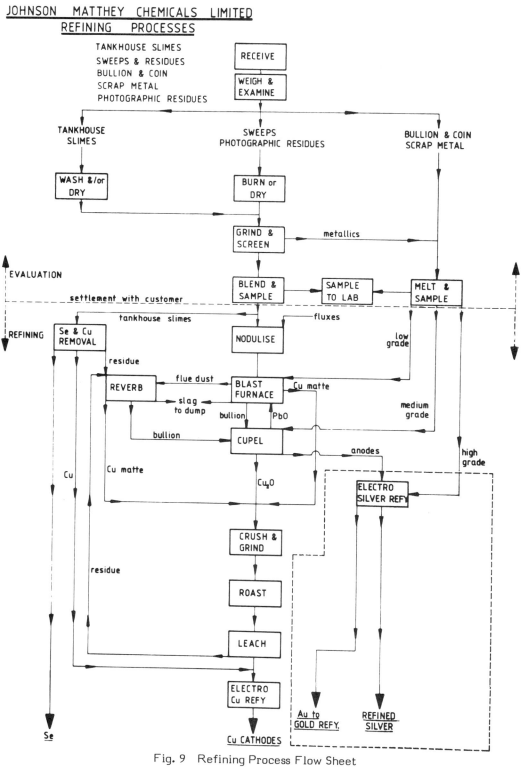

Fig. 9   Refining Process Flow Sheet

microprocessor controlled instrument can rapidly and accurately determine the composition of a solution, detecting up to 37 elements in only 2 minutes.

Results are fed to the stock control computer and then reported to the customer for acceptance.

Once this acceptance is received, materials are released for refining.

As can be seen from the flow sheet, three main material refining routes are followed:-

a)   Tankhouse Slimes

b)   Sweeps

c)   Bullion

Tankhouse Slimes:

A new process for the treatment of copper tankhouse slimes provides for the refining of selenium at high rates of recovery. The plant has several unique features, not least being its operation by computerised control. Selenium is first leached from the slime, precipitated and refined before grinding to the required mesh size. The residue is then further leached with dilute suphuric acid to remove copper for electrowinning. The final residue containing gold, silver and lead then joins the main smelting stream for recovery.

Fig. 10   Tankhouse Slimes Plant

Sweeps:

Charges for the blast furnace are calculated to give optimum amounts of gold, silver, copper, iron oxide, lime, silica, alumina prior to nodulising.

The payload materials are first blended with the requisite flux additions then fed to a disc pelletiser. With the addition of water, nodules are formed. These are dried and stored ready for smelting.

The blast furnace, with its forehearth and ancillary equipment, has been made to JMC's design. It has a hearth area of 23 sq.ft. and is 14ft. 9ins. high to charge level.

The furnace is charged by an inclined belt conveyor which in turn is fed from 14 hoppers controlled by mini processor.

Oxygen-enriched air is blown through 16 tuyeres located in mild steel water jackets forming the bosch of the furnace. The main constituents of the charge are coke, litharge, nodules (powder and fluxes), iron pyrites and reverberatory slag.

As the charge descends in the furnace, the fluxes and gangue melt forming slag, the litharge is reduced to lead which collects the gold and silver. The pyrites decompose to form ferrous sulphide and sulphur, which in turn converts any copper present into copper sulphide or matte.

The complete molten charge is tapped via an uplift tap hole into the forehearth which is designed so that the three phases produced in the blast furnace can be separated and tapped according to their density.

Molten slag flows continously from the top of the settler into a continuous casting machine. After analysis to ensure that it contains no values it is either sold as hardcore or dumped.

The second phase, copper matte, is tapped into ladles prior to casting on a second casting machine.

As well as copper, the matte contains quantities of gold, silver and lead.

After crushing and grinding the matte is roasted to copper oxide in a six hearth Herreshoff furnace manufactured by Neptune Nicholls. The roasted matte is leached with dilute sulphuric acid to remove the copper which is electrolysed to cathode. The residue, along with residue from the tankhouse slimes process, is smelted in a Monometer short bodied rotary reverberatory furnace with suitable fluxes and reducing agent to produce a rich lead bullion and slag. These are tapped into a Masters inclined casting machine.

The lead bullion which contains most of the gold and silver, is tapped from the forehearth into a ladle, before being cast into bars or poured directly into a cupellation furnace.

A cupellation furnace is basically a very shallow hearthed reverberatory furnace fired either by oil or gas.

The lead bullion, containing approximately 0.5% gold and 25% silver, is melted and air or oxygen-enriched air is blown onto the surface of the molten bullion. Lead, along with other base metals, is oxidised to litharge which floats to the surface as a thin molten layer and is run off into moulds. More lead bullion is added to the furnace until the bath contains only gold and silver with small amounts of copper.

This metal, assaying 2% gold and 96% silver, is cast into anodes which are electrolysed using the Moebius process.

All gases from the smelting and cupellation furnaces are handled by two 60,000 cfm Luhr bag filters which discharge cleaned gases into a 250 ft. stack provided with gas sampling points.

Flue dust is collected in sealed containers, nodulised and smelted in a rotary reverberatory furnace.

Bullion:

Bullion >98% silver is cast directly into anodes and electrolysed.

Bullion >70%. Bullion with silver content between 70% and 98% is refined by cupellation. The same process is used as was described for the treatment of lead bullion except that cuprous oxide is produced instead of litharge. This oxide, which contains approximately 22% silver, is treated by roasting and leaching processes, the copper being electrolysed to cathode and the silver residue melted into anodes for electrolysis.

F.T. Embleton

Fig. 11   Luhr Fume Extraction Plant

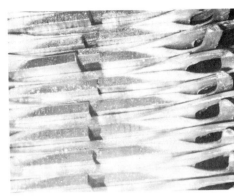

Fig. 12   Silver Anodes

Bullion < 70%. This metal is treated by first shattering (a process similar to atomising) and then leaching with sulphuric acid. The copper is electrowon and the resultant silver residue is melted and cast into anodes weighing approximately 800 ounces troy.

Electro Silver Refining, Moebius Process

In this process, the silver anodes containing gold are placed in ploypropylene bags and hung in a cell containing acid silver nitrate electrolyte.

A current of about 500 amps and 1.5/2.0 volts is passed.

Silver crystals of at least 99.9% purity are deposited on stainless steel cathodes.

A scraping mechanism continuously removes the silver from the cathodes, dropping it into boxes located at the bottom of the cells from which it is removed daily, washed, and then either melted into market bars or used to manufacture silver nitrate or other silver salts.

Any gold present in the anode is collected in the bag as a slime.

< 20% gold is fed into the cupellation furnace used for refining lead bullion.

> 20% gold is refined using the Miller Chlorine process which deals with gold alloys and scrap containing down to 20% but more usually 50% of gold upwards.

It is the most widely used and most important of all gold refining methods, depending on the fact that gold and platinum metals do not react with chlorine at the temperature of molten gold (1063 degrees C) whilst most base metals form molten or volatile chlorides under these conditions. The crude gold accumulated from high grade scrap and from earlier stages of refining is melted in clay crucibles and gaseous chlorine is passed into the bottom of the melt through a clay pipe. The chlorides of $ZnCl_2$, CuCl, $NiCl_2$ etc. together with AgCl, rise to the surface of the pot as a molten mass or as volatile chlorides. These slags can be removed during the chlorination stage by bailing off the surface and adding more crude gold to the pot. By this means, an increasing amount of gold can be accumulated in the crucible.

At the end of the reaction, the surface is finally cleaned up by the addition of salt flux prior to transferring the concentrated gold into a holding furnace. Usually, several small chlorination reactors are used and the end product is fed to a large tilting furnace for final casting of a uniform bulk of metal. The quality of gold obtainable by this process is normally 99.5% to 99.6% gold which is acceptable as "good delivery" gold bars on the world bullion markets.

Whilst higher quality gold can be obtained by this process, the time factor for chlorination is considerably lengthened and increasing losses of gold into the fume system takes place.

Fig. 13   Removing Silver Crystal

Fig. 14   Market Bars Stacked

All fumes from this process must pass into a special fume system, capable of keeping the working atmosphere around the furnace clear, and incorporating an electrostatic precipitator to ensure complete removal of values from the final gas stream.

A typical dust collected can contain up to 200 ounces troy of gold and 3,500 ounces troy of silver per ton to be fed back for recovery by pyrometallurgy. Any gold obtained by chlorination will still carry with it any platinum group metals and requires further refining to separate these metals.

Furthermore, with increasing use of gold and its alloys for specialised industrial applications, the presence of any impurities must be reduced to the minimum. Good delivery bars are not always suitable for industrial use and pure gold of 999 and 9999 quality is in large demand.

Pure Gold - High Quality

Our method of preparing pure gold depends upon the impurities present. If platinum group metals are present, the HCl/Cl_2 process is used; if no platinum group metals are present, the Wohlwill electrolytic process is usually used.

Chlorine/HCl Process

This process is used for refining slimes from the Meobius circuit and also gold from the Miller process which contains platinum group metals.

The gold from chlorination refining is first converted to grain by melting and pouring into water. The grain provides a large surface area for acid attack. The gold is dissolved in Cl_2/HCl in glass columns producing a gold chloride solution.

The gold can be readily precipitated by addition of SO_2 or sodium sulphite to the cold diluted solution and, under controlled conditions, a course granular precipitate is obtained which can be readily washed with water and gold of the highest purity obtained. This product, when dried, is fed into crucibles and melted. Gold of 99.99% purity is consistently obtained.

Fig. 15   Graining Gold

Fig. 16   Gold 9999 Bars

Wohlwill Electrolytic Process

Non-platiniferous gold is upgraded from the 99.6% chlorine refined gold by elecrolysis using the Wohlwill process.

The impure gold is cast into anodes and electrolysed in a gold chloride/HCl electrolyte. The cells operate at approximately 60 degrees at a voltage of 1.1/2.2 volts. The 99.9% gold is deposited onto Titanium cathodes. After removal from the cathode, the gold is melted into bars.

Because of the HCl/chlorine fumes produced, the whole plant is hooded and connected to an extract system.

For economic reasons, the plant is operated on a 24 hour basis with constant current control to allow for any increase of resistance due to anode dissolution. The plant must also operate on a continuous throughput as any shutdown results in interest payments on the gold locked up in the electrolyte.

Gold Recovery by Reduction of Solvent-Extracted Au (III) Chloride Complex — A Kinetic Study

G. F. Reynolds and Stephen G. Baranyai[1], Department of Chemistry and Chemical Engineering, Michigan Technological University, Houghton, MI 49931

and

Leonard O. Moore, Ansul Co., Technical Center
P.O. Drawer 1165, Weslaco, Texas 78596

ABSTRACT

Previous work has shown that gold can be separated commercially from other metals by dissolution in aqua regia, followed by extraction with diethylene glycol dibutyl ether and the subsequent reduction of the gold (III) complex. The rate of reduction of the complex has been quantitatively measured with a variety of aqueous phase reducing agents by following the disappearance of the characteristic 242 nm UV band of the gold (III) chloride-ether complex by periodic sampling of the organic phase. The reduction reaction has been found to be first order in both the gold (III) complex and the aqueous reducing agent, giving a second order overall reaction order. Under identical conditions of heating and stirring, it was found that hydrogen peroxide is the fastest reducing agent followed by hydrazine hydrate, sodium nitrite, formic acid, oxalic acid and sodium oxalate. A mechanism consistent with the data has been proposed.

INTRODUCTION

Previous work has shown that gold (III) chloride may be extracted from aqueous hydrochloric acid solutions by use of a wide variety of basic organic solvents.[2,3] Diethylene glycol dibutyl ether has been found to be a choice solvent for this extraction because it selectively removes gold (III) tetrachloride without extracting the platinum metals. The distribution coefficient of this extraction is in the region of 2500 for concentrations encountered in commercial processes.[4]

Gold (III) tetrachloride has been found to show a characteristic UV band at 245 nm in aqua regia and at 252 nm in diethylene glycol dibutyl ether.[5] This information shows that the extracted species is the same species that is present in aqua regia. This neutral species is the hydrated proton and the square planar gold (III) tetrachloride anion. The selectivity of the diethylene glycol dibutyl ether for the gold (III) tetrachloride has been attributed to the low charge to size ratio of the hydrated proton-gold (III) complex.[4]

Diethylene glycol dibutyl ether has some very desirable properties for use in industry. It has a boiling point of 245.6°C, a high flash point of 118°C, and is almost insoluble in water (0.3 w% at 20°C).[6]

In the method described by Morris and Ali Khan, gold (III) was extracted into diethylene glycol dibutyl ether from a precious metal concentrate (5% gold and 70%

other precious metals) in aqua regia. The "loaded" diethylene glycol dibutyl
ether was separated from the aqua regia solution and the latter was discarded.
The "loaded" organic solution was then washed with a 1.5 $\underline{M}$ hydrochloric acid solution to remove any base metals that could be present in the solvent. In the final
step, gold metal was recovered by heating the "loaded" organic solution with a 5%
solution of aqueous oxalic acid.

The kinetics of the reductive step in the aforementioned method are studied in
this work. In looking at the overall reaction equation:

$$2HAuCl_4(DGDE) + 3H_2C_2O_4(aq) \underset{\Delta}{\rightleftharpoons} 2Au(s) + 6CO_2(g) + 8HCl(aq)$$

gold (III) is reduced to gold metal and oxalic acid is oxidized to carbon dioxide.
In the Morris and Ali Khan procedure, the 5% aqueous oxalic acid solution represents a large excess of the reducing agent over the $HAuCl_4$ (auric acid). As the
reduction of gold (III) to gold metal occurs, the aqueous oxalic acid is in such
a large excess that it can be seen as remaining at a constant concentration. With
this assumption, the reaction order with respect to the gold (III) tetrachloride
can be determined.

If a stoichiometric concentration of aqueous oxalic acid is used instead of an excess, the reaction order of the overall reaction can be determined. From the difference between the overall reaction order and the pseudo-order of the gold (III)
tetrachloride, the reaction order with respect to the oxalic acid concentration
can be determined.

Rimmer states that other reducing agents (hydroquinone and sulfur dioxide) have
been used besides oxalic acid. In this work, formic acid, hydrogen peroxide,
sodium nitrite, hydrazine hydrate and sodium oxalate were tested as reducing agents
in place of oxalic acid. These reducing agents were selected because they have
favorable E° values for the reduction of gold (III) tetrachloride to gold metal.
Table I shows the E° values for the oxidation of some of the reducing agents.

Table I

E° Values of the Oxidation of the Reducing Agents[10]

Reducing Agent	E° Value
oxalic acid	0.49 v
formic acid	0.2 v
sodium nitrite	−0.88 v
hydrogen peroxide	−0.682v

EXPERIMENTAL

The following method was used to prepare the gold (III)-diethylene glycol dibutyl
ether complex: 1.0 ml of auric acid and 2.0 ml of diethylene glycol dibutyl ether
were pipetted into a separatory funnel. The funnel was shaken for two minutes.
After allowing the layers to separate, the aqueous layer was drained off and retained. The complex was emptied into a 25 ml Erlenmeyer flask and saved. The
aqueous layer was then poured back into the separatory funnel and 2.0 ml of fresh
diethylene glycol dibutyl ether were pipetted into the funnel. The mixture was
shaken for two minutes and the layers allowed to separate. The aqueous layer was
poured off and discarded. The remaining diethylene glycol dibutyl ether was added to the 25 ml Erlenmeyer flask and mixed gently. This resulted in a solution
that was 0.01270 $\underline{M}$ auric acid in diethylene glycol dibutyl ether.

The 25 ml Erlenmeyer flask with the "loaded" diethylene glycol dibutyl ether was placed in a water bath which was maintained at 68°C (± 0.5°). Then 4.0 ml of the 0.5 $\underline{M}$ reducing agent were pipetted into another flask. This flask was also set in the water bath. Approximately fifteen minutes were allowed for both solutions to obtain thermal equilibrium. The first sample was removed, with a 1 μl pipet, from the diethylene glycol dibutyl ether at this point. The 4.0 ml of reducing agent were poured into the flask with the complex and the timer was started, with stirring. Successive 1 μl samples were removed from the diethylene glycol dibutyl ether layer every 100 seconds. These samples were placed into separate vials and were then diluted with 2.0 ml of anhydrous methanol. Sampling was continued until the transfer of the gold (III) from the diethylene glycol dibutyl ether was nearly complete. This was indicated by the loss of the yellow color in the diethylene glycol dibutyl ether. The stoichiometric reduction was performed in the same manner except that the 0.02 $\underline{M}$ solution of the reducing agent was used.

A Beer's Law plot of gold (III) in diethylene glycol dibutyl ether was prepared by making successive dilutions of 0.02370 $\underline{M}$ HAuCl$_4$ in diethylene glycol dibutyl ether and removing a 1 μl sample from each dilution. This sample was then diluted with 2.0 ml of anhydrous methanol. A reference sample was prepared in the same manner with pure diethylene glycol dibutyl ether. Absorbance measurements were made at 242 nm of each of the samples on a Beckman DU-2 spectrophotometer in 1.0 cm silica cells. The absorbance values were then plotted versus the corresponding concentration of gold (III) tetrachloride in the diethylene glycol dibutyl ether.

To determine the gold (III) tetrachloride concentration in the samples from the reduction methods, the absorbance of each sample was measured on the Beckman DU-2. These values were then converted to gold (III) tetrachloride concentration by using the Beer's Law plot. Once the gold (III) tetrachloride concentrations were determined, the resulting values were plotted (as natural logarithmic functions) versus time, using a least squares program.

Oxalic acid was used as the reducing agent for the mechanism study. To begin the study, the role of water in the reductive step was examined. Finely ground oxalic acid was dried overnight in a vacuum oven at 40°C. A saturated solution of oxalic acid was prepared by dissolving 2.833 grams of oxalic acid in 25.0 ml of diethylene glycol dibutyl ether. Then 2.0 ml of this saturated solution was mixed (at 70°C) with 2.0 ml of the gold (III)-diethylene glycol dibutyl ether complex. No reaction was observed. As soon as a second aqueous phase was established by addition of 4 ml of distilled water, the reaction ran to completion.

RESULTS AND DISCUSSION

Both qualitative and quantitative experiments were carried out in order to determine the rate at which gold reduction occurred with the various reducing agents. By observing the decrease of the concentration of the gold chloride complex in the ether layer with time under analogous conditions of heating and stirring qualitative comparisons were made of the various reducing agents. The results shown in Table II indicate that when the various reducing agents are used in excess as 5% solutions, hydrogen peroxide is the most effective reducing agent followed by hydrazine hydrate, sodium nitrite, formic acid, and oxalic acid. Sodium oxalate effects little or no reduction upon heating to 80°, but preliminary experiments show that the addition of the phase transfer catalyst, dibenzo-18-crown-6-ether, can markedly increase the reduction rate.

Table II

Reducing Agent Effectiveness[a]

Reagent	Concentration	Time, hrs.	Product
oxalic acid	5%	3-5	gold sand
formic acid	5%	2-2.5	gold flakes
sodium nitrite	5%	1.5-2	gold flakes[b]
hydrazine hydrate	5%	1-1.5	black flakes
hydrogen peroxide	5%	0.5-1	gold sponge
sodium oxalate	5%	no rxn	----------

[a]The reducing agents are in an aqueous solution and the reductions are carried out between 70° and 80°C.

[b]On ignition to constant weight, these flakes became gold colored.

Quantitative kinetic runs as described in the experimental section were performed using hydrogen peroxide, oxalic acid and formic acid as reducing agents. By using these reducing agents in large excess, psuedo first order kinetics were followed in each case. Good linear plots were obtained when the natural logarithm of the gold concentration remaining in the ether layer were plotted against time. Table III shows the pseudo first order rate constants obtained for reduction with various concentrations of oxalic acid and formic acid at 68°C.

Table III

Pseudo-Order Reduction Results at 68°C.

Reducing Agent	Conc'n	$[HAuCl_4]$	$k_1 \times 10^3$ sec^{-1}
oxalic acid	0.6071	0.01270	3.49 ± 0.23
oxalic acid	0.6071	0.01270	2.43 ± 0.07
oxalic acid	0.2428	0.01270	4.62 ± 2.10
oxalic acid	0.2428	0.02540	2.45 ± 0.19
formic acid	1.060	0.01270	1.30 ± 0.24
formic acid	1.060	0.01270	1.28 ± 0.08
formic acid	1.325	0.01270	2.02 ± 0.30
formic acid	1.325	0.01270	2.15 ± 0.14

When neither the reducing agent nor the gold chloride complex were in excess, the reaction rate was found to follow second order kinetics. For oxalic acid and hydrogen peroxide, second order rate constants for the reactions having the stoichiometry A + 3/2 B → products can be obtained from the following equation:[7,8]

$$\frac{1}{\frac{3}{2} a - b} \ln \frac{b(a - x)}{a(b - \frac{3}{2} x)} = k_2 t,$$

where a and b are the initial concentrations of A and B, respectively, and a - x and b - x are the concentrations of A and B, respectively, remaining at time t. In our case, A is the gold chloride complex and B is the reducing agent oxalic acid or hydrogen peroxide. The second order rate constants obtained using oxalic acid and hydrogen peroxide are shown in Tables IV and V.

Table IV

Stoichiometric Reduction Results at 68°C.

Reducing Agent	Conc'n	[HAuCl$_4$]	$k_2 \dfrac{\ell}{mol\ sec}$
oxalic acid	0.01897	0.01270	0.951 ± 0.082
oxalic acid	0.01897	0.01270	0.424 ± 0.061
oxalic acid	0.01897	0.01270	0.925 ± 0.130
oxalic acid	0.01897	0.01270	1.16 ± 0.216
oxalic acid	0.01897	0.01270	0.496 ± 0.058
hydrogen peroxide	0.0286	0.01270	0.126 ± 0.046

Table V

Stoichiometric Reduction Results at 55°C.

Reducing Agent	Conc'n	[HAuCl$_4$]	$k_2 \dfrac{\ell}{mol\ sec}$
hydrogen peroxide	0.0286	0.01270	0.064 ± 0.003
hydrogen peroxide	0.0286	0.01270	0.061 ± 0.007

The rate constants for the stoichiometric reduction average $0.791 + 0.28 \dfrac{\ell}{mol\ sec}$ for oxalic acid, $0.062 + 0.002 \dfrac{\ell}{mol\ sec}$ for hydrogen peroxide at 55°C and $0.126 + 0.05 \dfrac{\ell}{mol\ sec}$ for hydrogen peroxide at 68°C. The energy of activation can be calculated from the equation:

$$E_a = R\frac{T_2 T_1}{T_2 - T_1} \ln \frac{k_2}{k_1}$$

in which R is the ideal gas constant, T_2 and T_1, the different temperatures for k_2 and k_1, respectively. Using this equation, the approximate energy of activation for hydrogen peroxide has been calculated to be about 12 Kcal/mole. The calculated value is in the same region as values calculated for other redox systems.[14]

The experimental results indicate that the reduction of gold (III) to gold metal using oxalic acid, formic acid and hydrogen peroxide, proceeds by a bimolecular mechanism. A bimolecular rate determining step is indicated by the fact that the reaction is first order in gold chloride complex and first order in the reducing agents studied. In proposing a possible mechanism, only oxalic acid will be considered to demonstrate the reaction path. The other reducing agents would be expected to follow an analogous mechanism.

In the mechanism study, it was found that without water present in the reduction system, no reaction was observed. But when water was added to the system, the reaction ran to completion. This result indicates that the reduction of gold (III) to gold metal took place in the aqueous layer or along the phase boundary between the organic and aqueous phases. To determine which conclusion was correct, the reduction experiment was run simultaneously in an NMR tube and a 50 ml Erlenmeyer flask. The reaction ran to completion in both vessels at the same rate. This would lead one to believe that the reduction takes place in the aqueous phase and not on the phase boundary.

The proposed mechanism for the reduction of gold (III) to gold metal using oxalic acid is shown in the scheme below. When aqueous oxalic acid is added to the "loaded" organic complex, there are two rapid equilibria set into motion. The first (Eqn 1), is a transfer of oxalic acid from the aqueous phase to the organic phase. The other equilibrium (Eqn 2), is a transfer of gold (III) from the organic phase to the aqueous phase. Once the gold (II) is in the aqueous phase, the substitution of oxalic acid into the square planar complex of the gold (III) occurs (Eqn 3). The oxalic acid is then oxidized to carbon dioxide (Eqn 4) and the gold (III) is then reduced to gold (I). The last step of the mechanism (Eqn 5) is the disproportionation of the gold (I) to gold metal and to gold (III).

$$H_2C_2O_4 (aq) \underset{}{\overset{K_1}{\rightleftharpoons}} H_2C_2O_4 (DGDE) \qquad \text{(Eqn 1)}$$

$$HAuCl_4 (DGDE) \underset{}{\overset{K_2}{\rightleftharpoons}} HAuCl_4 (aq) \qquad \text{(Eqn 2)}$$

$$HAuCl_4 (aq) + H_2C_2O_4 (aq) \underset{slow}{\overset{k_1}{\longrightarrow}} HAu(HC_2O_4)Cl_3 (aq) + HCl (aq) \qquad \text{(Eqn 3)}$$

$$HAu(HC_2O_4)Cl_3 (aq) \underset{fast}{\overset{k_2}{\longrightarrow}} HAuCl_2 (aq) + 2CO_2 (g) + HCl (aq) \qquad \text{(Eqn 4)}$$

$$3HAuCl_2 (aq) \underset{fast}{\overset{k_3}{\longrightarrow}} 2Au(s) + HAuCl_4 (aq) + 2HCl (aq) \qquad \text{(Eqn 5)}$$

The equilibrium (Eqn 1) of the oxalic acid between the aqueous phase and the organic phase has been shown to exist not only by the solubility of the oxalic acid in the diethylene glycol dibutyl ether (11.3 g/l), but by the strong UV absorption at 265 nm in diethylene glycol dibutyl ether after 100 seconds of mixing at 68°C. The UV absorptions for oxalic acid reported in the literature are a peak at 280 nm and a shoulder at 250 nm.[9,10] The difference between the observed wavelength and the literature wavelength can be explained by the solvent shifting of the observed wavelength.

The equilibrium represented in Equation 2 is readily observed by the transfer of the yellow colored gold (III) species from the diethylene glycol dibutyl ether to the aqueous phase and from the characteristic UV absorption at 252 nm of the gold (III) in the aqueous phase. The cause of the displacement of the gold (III) from the organic to the aqueous phase is believed to be due to the equilibrium of oxalic acid between the two phases. The oxalic acid could be considered to be a phase transfer reagent for this step.

Equation 3 is the substitution (in the aqueous phase) of the oxalic acid into the gold (III) square planar complex. This is proposed to be the rate determining step of the mechanism. The substitution of the ligand into square planar complexes has been well studied.[11,12] This substitution follows a bimolecular displacement path (S_N2) and is often the first step in a redox process.[13] The first step of the substitution is the approach of the oxalic acid toward the gold (III) complex. The oxalic acid attacks the gold (III) complex and forms a trigonal bipyramid intermediate. This is then converted back into the square planar gold (III) by ejection of a chloride ion.

This substitution pathway is supported by at least four different kinds of experimental evidence.[14] The evidence is: (a) the fact that the substitution occurs with a retention of configuration; (b) the dependence of the rate of reaction on

the entering ligand; (c) the steric effects of substituents on the rates; and
(d) the isolation of many five- and six-coordinated complexes.

In this work, only the second piece of evidence is observed. The other three are
not of great importance to this mechanism. The dependence of the rate of reaction
on the entering ligand is shown in Table 1. The smaller sized molecules (hydrogen peroxide) react at a much faster rate than the larger molecules (oxalic acid).

Once the oxalic acid is substituted into the square planar complex, it is oxidized
to carbon dioxide as the gold (III) is reduced to gold (I). This is shown below:

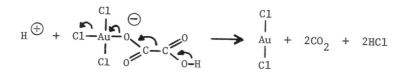

This redox step can be considered rapid as compared to the substitution step. The
gold (I) species formed from the redox reaction is unstable in water and rapidly
disproportionates to yield gold metal and gold (III).[15] This is seen in Equation
5.

In defining the rate of disappearance of the gold (III) from the diethylene glycol
dibutyl ether, the equilibrium represented in Equation 2 can be written as:

$$K_2 = \frac{[\text{HAuCl}_4(\text{aq})]}{[\text{HAuCl}_4(\text{DGDE})]} \qquad (\text{Eqn 6}),$$

or

$$[\text{HAuCl}_4(\text{aq})] = K_2 [\text{HAuCl}_4(\text{DGDE})] \qquad (\text{Eqn 7}).$$

The rate expression for the substitution step can be written:

$$\frac{-d[\text{HAuCl}_4(\text{aq})]}{dt} = k_1 [\text{HAuCl}_4(\text{aq})][\text{H}_2\text{C}_2\text{O}_4] \qquad (\text{Eqn 8}).$$

By substituting Equation 7 in for $\text{HAuCl}_4(\text{aq})$ in Equation 8, it is seen that:

$$\frac{-d\ \text{HAuCl}_4(\text{aq})}{dt} = \frac{-d\ \text{HAuCl}_4(\text{DGDE})}{dt} = K_2 k_1 [\text{HAuCl}_4][\text{H}_2\text{C}_2\text{O}_4] \quad (\text{Eqn 9}).$$

Therefore, by measuring the gold (III) concentration remaining in the diethylene
glycol dibutyl ether during the substitution reaction (the rate determining step)
as was performed in this study, the combined second order rate constant, $K_2 k_1$ is
determined. This combined rate constant is the value given in Table III for the
stoichiometric reductions. Thus, the above mechanism proposed for the reductions
is consistent with the kinetics observed when the aqueous phase reducing agents
are added to the gold (III) chloride complex in diethylene glycol dibutyl ether.

ACKNOWLEDGEMENTS

The authors wish to thank the Ansul Company of Marinette, Wisconsin for their
support of this study.

REFERENCES

1. Present address of S. G. Baranyai, Jr: Texas Instruments Inc., P.O. Box 225012 M/S 54, Dallas, Texas 75265.
2. E. B. Sandell, Colorimetric Determination of Traces of Metals, 3rd. ed., Interscinece Publishers, Inc., New York, 1959.
3. G. H. Morrison and H. Frieser, Solvent Extraction in Analytical Chemistry, J. Wiley & Sons, Inc., New York, 1957.
4. D. F. C. Morrison and M. Ali Khan, Talanta, 15, 1301-1305 (1968).
5. H. C. Serggeant and N. M Rice, The Mechanism of Uptake of Gold by Dibutyl Carbitol, The University of Leeds, U.K., 1977.
6. B. F. Rimmer, Chemistry and Industry, 2, 63-66 (1974).
7. R. G. Wilkins, The Study of Kinetics and Mechanism of Reactions of Transition Metal Complexes, Allyn and Bacon, Inc., Boston, 1974, Ch. 1.
8. W. J. Moore, Physical Chemistry, 4th ed., Prentice-Hall, Inc., Englewood Cliffs, N.J., 1972, pp. 334-336.
9. R. Wright, J. Chem. Soc., 103, 528 (1913).
10. CRC, Handbok of Chemistry and Physics, 54th. ed., CRC Press, 1973, pp. D-120, C-403.
11. R. G. Wilkins, op. cit., p. 184.
12. A. G. Sykes, Kinetics of Inorganic Reactions, Pergamon Press, Oxford, U.K., 1966, Ch. 13.
13. E. Chaffee and J. O. Edwards, "Replacement as a Prerequisite to Redox Process," Progress in Inorganic Chemistry, Vol 13, Inorganic Reaction Mechanisms, ed. J. O. Edwards, Interscience Publishers, 1970, pp. 205-242.
14. F. Baslo and R. G. Pearson, Mechanisms of Inorganic Reactions, 2nd. ed., Wiley & Sons, Inc., New York, 1967, pp. 375-380, 497.
15. E. M. Wise, op. cit., p. 41.

Oxygen Refining of Smelted Silver Residues

T.S. Sanmiya, R.R. Matthews

INCO Metals Company, Copper Refinery,
Copper Cliff, Ontario, Canada

ABSTRACT

Silver bearing slimes from copper electrorefining have traditionally been sulphated, smelted in a Dore' furnace and refined to Dore' bullion with the use of nitre. This refining method generates undesirable nitrogen dioxide emissions and substantial quantities of silver and precious metal bearing slag.

The process described, which was adopted by INCO Metals Company in 1974 at their Copper Refinery, provides an attractive alternative to the use of nitre with improved environmental conditions, less slag production, and improved distribution of selenium and tellurium for subsequent recovery. Oxygen injection followed by two fluxing steps produces Dore' silver bullion suitable for electrolytic refining.

KEYWORDS

Oxygen; refining; silver; Dore' furnace; fluxing; slags; slimes; NO_x emissions; environment;

INTRODUCTION

INCO Metals Company, along with many other copper producers, has used nitre as an oxidant at some stage of their respective anode slimes treatment processes, with the practice still being employed at some refineries. In 1969, a decision was made by INCO to replace the nitre refining method and eliminate the characteristic red brown plume caused by the emission of nitrogen oxides (NO_x). This became a reality in 1974.

To put the oxygen refining process in perspective, a brief outline of the Copper Refinery and Silver Building flowsheets, and the present Dore' furnace operation precede the description of the oxygen refining process.[1] An extensive program lasting four years, until the commissioning of the new process, is discussed with the remainder of the paper focusing on details and results of the oxygen refining procedure as initially implemented. Numerous refinements, gained from the past 6 years of operating experience, have been made to the original process.

---

[1]Canadian Patent No. 981,911

101

Copper Cliff Copper Refinery

The Copper Refinery is situated about one mile from the smelter and receives molten copper in refractory lined hot metal cars. Figure 1, shows a simple schematic diagram of the INCO Copper Refinery.

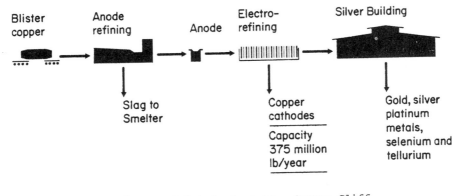

Fig. 1.   INCO Metals Company - Copper Cliff
Copper Refinery (CCCR).

Three 400 ton reverberatory anode furnaces upgrade the metal to about 99% copper. Cast anodes are electrorefined in the Tankhouse which has 1,560 cells and a plating capacity of 375 million pounds per year of high quality cathode copper. Anode slimes produced at the rate of 8 to 10 pounds per ton of cathodes are collected and processed at the on site Silver Building.

Silver Building

The silver building is divided into four principal areas consisting of slimes sulphation for copper and nickel removal; furnace smelting and refining to produce a Dore' bullion suitable for electrorefining; selenium and tellurium recovery; and finally the silver parting room and associated gold and platinum group metals recovery facilities. The schematic diagram in Fig. 2, shows the relation of the Dore' furnace to the remainder of the Silver Building processes. Anode slimes are thickened, filtered and processed in a reactor to sulphate copper and nickel. Leached, sulphated slimes are then smelted with suitable fluxes in a Dore' furnace to produce Dore' metal containing about 50% silver. The furnace bath is upgraded by oxygen refining to above 90% silver and less than 1% copper, making the Dore' bullion suitable for electrolytic refining of silver, analyzing 999.8 fine or greater. The subsequent silver parting slimes are treated by hydrometallurgical techniques to produce fine gold and platinum metal group concentrates.

Dore' Furnace Operation

The Dore' furnace is a small reverberatory type, (Fig. 3) having a single gas fired burner in the roof and with a capacity of about 30,000 pounds of Dore' metal. Leached, sulphated slimes are charged as a mixture containing suitable fluxes and reverts with sufficient fusion time provided before the next charge. This procedure continues until the furnace is full, at which time slag skimming begins. Alternate skimming and charging is continued until the furnace is full of Dore' metal.

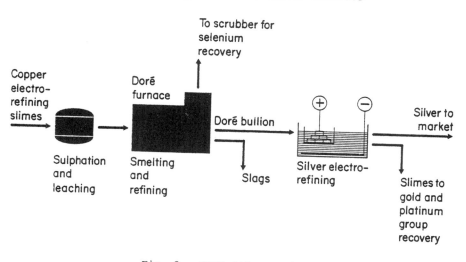

Fig. 2.   CCCR Silver Building

The smelting slag, having significant lead content, is reverted to the smelter complex for subsequent treatment.

Traditionally, the refining of Dore' metal at the INCO Copper Refinery was done by injection of nitre with air using a fettling machine during the entire refining period. This produced about 16,000 pounds of nitre slag per typical Dore'campaign, which had to be treated for both selenium and tellurium recovery. The final 4,000 pounds of slag was given the nomenclature of 'special nitre slag', which required separate treatment because of exceptionally high platinum group metals concentrations.

Fig. 3.   Reverberatory Dore' furnace.

Currently, the Dore' metal is subjected to the oxygen refining process, described in detail later in the paper, which upgrades the silver content to greater than 90% in the Dore' bullion. Virtually all of the selenium in the Dore' metal is collected in the flues and scrubbing system. Slags produced are oxide slag, which concentrates residual lead and nickel and a substantial portion of the copper; soda slag, which primarily collects tellurium; and borosilicate slag which accomplishes final copper removal to less than the 1%.

The Dore' bullion is cast as thin, 25 pound, anode plates which are electrorefined in silver parting cells.

DEVELOPMENT

Laboratory Scale

Preliminary tests, consisting of two pound charges of Dore' metal, were conducted starting in 1969 to investigate possible alternative oxidants for nitre. Tests were restricted to those oxidizing materials which were readily available and relatively inexpensive. Sodium peroxide, sodium borate, sodium carbonate and oxygen enriched air were tested. Besides elimination of NO_x emissions, other considerations were acceptable Dore' bullion, costs, amounts and types of slag, workroom environment, materials handling difficulties, refractory attack and disposal of sodium rich waste resulting from slag treatment. By itself, solid fluxing was considered unsuccessful for the following reasons:

1.   The silver content of the Dore' bullion did not exceed 82%.

2.   Crucible attack was severe with some mixtures.

3.   Slag viscosity was often unsuitable for good contact with the Dore' metal and reaction rates were therefore severely compromised.

Air injection tests were conducted with 25% to 50% oxygen enrichment which provided the required mixing in contrast to solid fluxing tests and in fact, special precautions were necessary to control excessive splashing. Initial oxygen requirements were estimated by assuming the production of CuO, NiO, SeO_2, TeO_2, and PbO. Various flow rates, processing times and stoichiometric oxygen multiples were tested with successful results near 2.5 times the stoichiometric volume of oxygen.

The initial laboratory tests were followed by scaled up tests with 50 pound charges of Dore' metal and 50% oxygen enriched air injection. Sodium carbonate was also introduced to simulate the fluxing by-product of nitre decomposition. Extraction of the base metals, selenium and tellurium was complete with the exception of copper which at best remained at a level of 2% in the Dore' bullion. Fluxing indicated that a mixture of three parts sodium borate and one part silica (borosilicate flux) would successfully lower the copper content below the 1% target while maintaining acceptable slag fluidity. The success of these tests prompted further work on a pilot scale.

Pilot Scale

A series of tests were conducted early in 1971, each with 4,000 pounds of Dore' metal, in an obsolete production furnace which had been refurbished for the tests. Air and sodium carbonate were used initially, which made it difficult to maintain a sufficiently high bath temperature and partial freeze-ups were frequent. Also, sodium carbonate usage was required throughout the refining period which produced

unacceptably high quantities of slag and silver recycle. This procedure successfully extracted the selenium, tellurium, lead and nickel but copper could not be
lowered below 3% in the Dore' bullion.

A further test with oxygen enriched air, and no flux addition, produced a similar
Dore' bullion but with a higher tellurium content. Borosilicate flux which was
tried for the first time on a large scale, collected the residual tellurium and
lowered the copper content of the bath to 1.0%. The pilot scale tests demonstrated
that refining with oxygen enriched air followed by a borosilicate fluxing step,
achieved similar performance to nitre refining with respect to the Dore' bullion
quality. Slag production and treatment was noticeably reduced and refractory attack was light. In view of these results and similar costs and process times compared to nitre refining experience, production scale testing was initiated.

Production Scale

Full scale tests, starting in 1972, were conducted on normal production batches of
about 27,000 pounds of Dore' metal to reaffirm and optimize the basic oxygen refining method. The silver concentration of the Dore' metal was successfully increased from 53% to near 96% producing typically 13,700 pounds of Dore' bullion
acceptable for electrorefining.

To accomplish this, the bath was first prepared by skimming clean all residual
slags and raising the bath temperature as required to 1,150°C. Commercial oxygen was used exclusively for all production tests having previously demonstrated
the technical feasibility and safety of its use during pilot testing.

Oxygen was injected below the bath surface through a stainless steel lance with its
tip bent as shown in Fig. 4, to permit suitable immersion.

Fig. 4. Oxygen injection.

The flow rate was adjusted within a range of 40 to 80 s.c.f.m. which provided sufficient but not excessive agitation, and controlled the exothermic reaction rate, primarily a result of selenium oxidation. Oxygen injection and skimming were alternated and towards the end of the process, slagging of impurities was assisted by the use of solid fluxing reagents which achieved the target impurity levels in the Dore' bullion. The tests indicated an oxygen consumption rate of 8,000 to 10,000 s.c.f. per ton of Dore' metal refined.

DISCUSSION

Refining

Dore' metal, at the beginning of the oxygen refining process, usually contains about 53% silver, 22% selenium, 8% copper, 7% tellurium, 6% lead and 2% nickel with the balance principally composed of gold and platinum group metals.

Figure 5, and Fig. 6, graphically illustrated the typical refining rate of Dore' metal during oxygen refining.

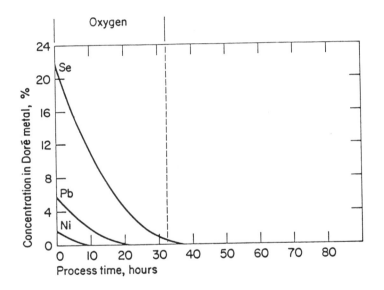

Fig. 5.   Refining rate - selenium, lead, and nickel.

For the initial 32 hours, oxygen injection does not require the use of any solid fluxing agents. Oxygen injection alone will remove the lead, nickel and selenium and reduce the copper and tellurium concentrations to 4% and 2% respectively. The silver concentration, during this period, increases from its initial 53% to greater than 85%.

The oxidation reactions are rapid during the early part of this period and the temperatures of the furnace bottom and bath usually increase despite the cooling effect of the injected oxygen. When the selenium concentration has been substantially reduced, however, the production of heat from exothermic reactions decreases, and some of the splashed metal often freezes. This solidified portion, containing appreciable concentrations of copper and tellurium, melts when solid fluxing begins. As a result, an increase in copper and tellurium concentration occurs in the bath which explains the apparent anomalies in the refining rate curves of Fig. 6.

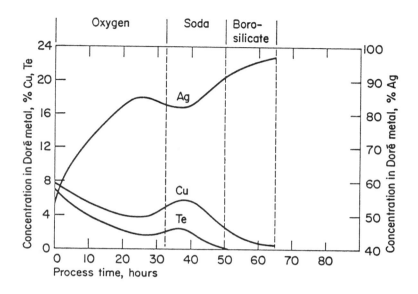

Fig. 6.   Refining rate - copper, tellurium and silver.

The existence of a silver - copper oxide eutectic at a copper concentration of 2% to 3% precludes completing the refining without the use of a solid fluxing agent. In addition, the removal of tellurium at low concentrations also becomes increasingly difficult when only oxygen is used.   Therefore, sodium carbonate is added in 500 pound increments accompanied by oxygen injection to collect the tellurium. Alternate skimming and soda fluxing is continued for about 18 hours or until the tellurium is reduced to less than 0.1%.   The copper concentration also decreases to about 3% but further reduction during soda fluxing becomes unacceptably inefficient.   To further reduce the copper content, 500 pound increments of borosilicate flux are used in conjunction with oxygen injection for a period of about 15 hours.   This reduces the residual copper to less than 1% in finished Dore' bullion now assaying 96% silver with the balance primarily gold and platinum group metals.

Slags

The oxide slag produced during the initial 32 hours of refining, totals about 4,500 pounds per typical Dore' campaign.   The slag is normally grey to black in colour, and varies from dry and thick to non-viscous depending on the impurity levels of selenium, lead and nickel in the Dore' metal.   It contains about 50% total copper, nickel and lead, and less than 10% total selenium and tellurium.

Soda slag, amounting to 1,500 pounds, is an easily skimmed, fluid slag having the characteristic dark, glassy appearance.   It contains about 35% tellurium and 10% copper which reflects its primary purpose to collect tellurium and reduce the copper content to less than 3% in the Dore' bullion.

The final borosilicate slag is also dark and glassy in appearance.   About 3,500 pounds is produced, containing about 25% copper and less than 5% of other major Dore' metal impurities.

All three types of slags have been found to be only slightly corrosive to the furnace refractory. Oxide slag which is the most severe, has a corrosion rate comparable to previous experience with nitre slag.

Slag Treatment

The oxygen refining process has substantially simplified slag treatment. A comparison of slag weights and assays for both oxygen and nitre refining is shown in Fig. 7.

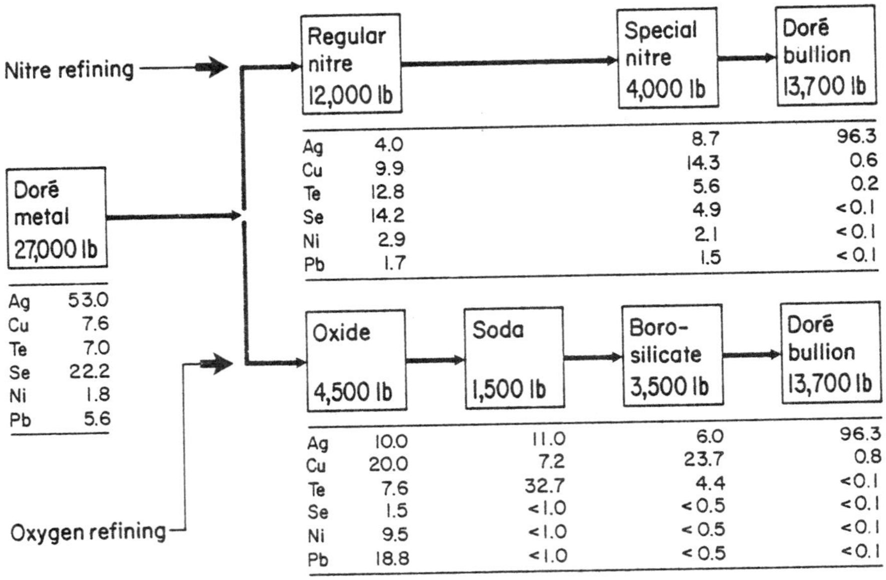

Fig. 7. Comparison of refining slags - assays in percent.

The entire 16,000 pounds of nitre slags produced per typical Dore' campaign required crushing, milling, roasting, leaching, and cementation operations to recover tellurium and selenium and to remove copper, before recycling the residue back to the Dore' furnace. In contrast, only about 6,000 pounds of the oxygen refining slags require recovery treatment which is generally less complex than methods used for nitre refining slags.

Since implementation of the oxygen refining process, oxide slag has been sulphated and leached to remove copper and nickel before recycle to the Dore' furnace. However, recent improvements to the process have substantially reduced the silver and tellurium content of this slag which now permits direct reverting to the anode furnaces. Soda slag is caustic leached for tellurium recovery with the residue recycled to the Dore' furnace. Borosilicate slag remains untreated and is ideal for fluxing the charging material to the subsequent Dore' campaign. About 95% of the silver and 3% of the copper in the borosilicate slag reports to the Dore' metal. The low selenium concentrations of the three oxygen refining slags precludes separate treatment of this element as was found necessary for nitre refining slags.

Metals Distribution

A comparison of the distribution of the primary recovered metals is shown in Table 1 for both oxygen and nitre refining.

TABLE 1 Typical Metals Distribution - % of Recovered Metals

	Ag	Cu	Ni	Se	Te	Pb
Nitre Refining						
Flue Solids	2	1	3	63	0	82
Regular Nitre Slag	3	54	81	33	90	14
Special Nitre Slag	3	38	16	4	8	4
Dore' Bullion	92	7	0	0	2	0
Oxygen Refining						
Flue Solids	2	1	2	99	50	17
Oxide Slag	3	56	98	1	15	83
Soda Slag	1	13	0	0	30	0
Borosilicate Slag	2	23	0	0	5	0
Dore' Bullion	92	7	0	0	0	0

The distributions of silver, copper and nickel are unchanged with the new process. Selenium, during oxygen refining, reports almost entirely to accessable areas of the flue recovery system compared to 60% for nitre refining. Thus, in the latter case, 40% of the selenium had to be recovered from the slag. Although oxygen refining distributes 50% of the tellurium to the flue system compared to nil for nitre refining, no detrimental effects on tellurium recovery results. The oxide slag tends to be an excellent lead collector with less than 20% reporting to the flue system compared to 80% for nitre refining.

Standard stack loss tests verified that emissions of all primary metals were virtually the same for both refining methods.

Lances

Standard lances for oxygen injection are made from 3/4 inch, schedule 40, stainless steel pipe. The 10 foot lance is attached to a flexable, armoured hose; the tip is bent into the bath and the lance is then secured in position using a simple clamp arrangement as shown in Fig. 8. As the lance tip burns, it is adjusted to ensure that a suitable immersion depth is maintained. The lance consumption rate varies from almost nil to 6 feet per hour, depending on the type of slag being produced. Normally, 8 to 12 lances are consumed when refining 27,000 pounds of Dore' metal.

Environment

With the adoption of the process in 1974, oxygen refining of the Dore' metal has completely eliminated the production of NO_x emissions. The characteristic red brown fumes are no longer emitted from the stacks.

The workroom environment has also been substantially improved. Blowing flux into the furnace using the fettling machine and high pressure air, is no longer required. In addition, recent innovations which changed the disposition of the slags has allowed removal of slag crushing and milling equipment.

An added benefit of the oxygen refining process is a reduction of 85% of the sodium which had been introduced during nitre refining. This has eliminated a major disposal problem for acid waste residue.

Fig. 8. Lance assembly

CONCLUSION

In this paper, the chronological steps taken to achieve the successful implementation of the oxygen refining process for smelted silver residues, have been outlined The major objective, to eliminate NO$_x$ emissions was achieved while providing an improved work environment, reduced slag production, and an improved metals distribution. These gains were achieved with comparable cost and precious metals recovery to the replaced nitre refining method.

ACKNOWLEDGEMENT

The authors acknowledge INCO Metals Company for permission to publish the information contained in this paper. Special thanks is extended to those at the Copper Refinery, whose valuable suggestions are greatly appreciated.

REFERENCES

Schloen, J.H., and E.M. Elkin (1954). Treatment of electrolytic copper refinery slimes. In Allison Butts (Ed.), Copper, The Science and Technology of the Metal, It's Alloys and Compounds. Reinhold Publishing Corporation, New York. pp 265 - 289.

Monahan, R.K., and F. Loewen (1972). Treatment of Anode Slimes at the INCO Copper Refinery. C.I.M. Conference.

Atmore, R.B., D.D. Howat, and P.R. Jochens (1971). The effects of slag and gold bullion composition on the removal of copper from mine bullion by oxygen injection. Journal of the South African Institute of Mining and Metallurgy, August, 5 - 11.

Parlee, N.A.D., and E.M. Sacris (1967). Oxygen - oxide equilibria in liquid silver - copper alloys. Transactions Of The Metallurgical Society of AIME. 239, 2005 - 2008.

Some Aspects of Precious Metals Recovery in the Jewellery Manufacturing Industry

Fernand Verschaeve
Chief Chemist
French Jewellery Company of Canada Limited
3750 Chesswood Drive,
Downsview, Ontario
M3J 2W6
Canada

ABSTRACT

Some three types of precious metal recovery in the manufacturing process of jewellery are described. Some thought has been given on the scientific approach versus the purely artistic one.

KEYWORDS

Scientist, jeweller, bombing, diamond cutting, deburring and tumbling end tanks.

INTRODUCTION

The purpose of this paper is to demonstrate that the classical battle between artistry and science is taking a different turn in the jewellery manufacturing industry.

The manufacturing of jewellery, seen through the eyes of a chemist, is full of bewilderment, disbelief and even magic. The logical thinker, the person who is trained to believe that all processes can somehow be defined, that all steps along the line can be tested in a cool, unemotional fashion, is in for a rude awakening. He is surrounded by hordes of skilled people who make a living for themselves by daily seemingly defying all scientific rules and regulations. These people do not only just get away with it, they thrive on it, even worse: they triumph. The scientist is ashen-faced to see a jeweller use his micrometer as a hammer, his thumb as a microhardness tester, his hand as a tensometer and his eyes as a UV-VIS spectrophotometer. The scientist is deeply saddened to learn that the jeweller can gauge the gold and silver content of an alloy long before he had a chance to start up his fire assay furnace or set up his burette. In a vein attempt to save face, the scientist approaches Top Management and asks for funds: funds for better equipment, funds for superior inspection devices, funds for a number of things. The scientist honestly wants to change the world of the artist - the jeweller - and turn it into a - what he thinks - a more cerebral and better one. For a while, the scientist prospers - he gets bolder and finally daringly proposes some form of Quality Control! This is where he meets his untimely end. He is now left with two choices: a) head for the CN Tower b) leave the artist-jeweller alone and try to recover some precious metal. Lacking in courage he chooses the latter avenue.

Fortunately for the chemist: the jewellery manufacturing industry offers a number of recovery sources. I shall very briefly touch on a few of them. I toyed for quite a while with the idea of showing some slides on gold and silver recovery, but I finally decided against it, for a number of reasons - the more outstanding reasons being lethargy and the rising cost of colour slide film in the wake of the decrease in the price of silver. Whatever: the recovery systems themselves are surprisingly simple and really don't need to be illustrated.

A first recovery system I would like to talk about is - what is called "recovery from bombing operations". "Bombing" is a rather emotional term: it conjures visions of big balls of fire and sounds of ear-deafening explosions. I should admit that I was rather hesitant myself the first time I was summoned to the "Bombing Area". In day-to-day jewellery manufacturing bombing is a rather sedate undertaking. All what really takes place is a chemical cleaning of chains and other jewellery products with an aqueous solution of potassium cyanide and hydrogen peroxide. The chains are dipped in this solution under a fumehood, until the sizzling more or less stops. Interestingly enough the "bomber" (= the person who does the bombing) keeps the ration $KCN - H_2O_2$ more or less a secret - why I don't know. Quite obviously the cyanide will dissolve some gold, silver and base metals at the surface of the chains. The precious metals in solution are not only easily recoverable, but recovering them is a far more rewarding experience than letting them go down the drain and into Lake Ontario. One way of getting all the gold out of solutions is to collect all bombing solutions and further water rinsings in a large tank which doesn't dissolve in cyanide - a collector tank or "right hand side tank" as we call it - then pumping everything through a special ion exchange resin column - into a treatment tank or "left hand side tank".

The amount of gold dissolved on a daily basis through bombing can accurately be gauged by taking a representative sample of the solution in the right hand side tank - providing the volume of the tank is known - and assaying it for gold. Methods which can be used are the simple gravimetric sulphuric acid method, if need be followed by cupellation and parting or an atomic absorption approach. The efficiency of the ion exchange resin absorption can be established by daily testing the solution of the left hand side tank too. The latter testing not only tells the chemist when the time is ripe to change the ion exchange resin load, but far more important, it convinces the bomber that the system really works. You have proudly shown him every day the gold bead you have recovered in the sample of the right hand side tank and you have impressed upon him that your sample, is only a fraction of the total volume: in other words that there is 500 even 1,000 times more gold in the collection tank than there is in your hand. All you have to do now is to make certain that virtually no gold comes through in the left hand side tank and he may start to believe that the gold is really present in the ion exchange column.

Taking the resin out is one easy thing, drying it and analyzing it for gold is another. You must make certain that all is done under controlled conditions, otherwise the restaurants in the neighbourhood of the plant will show a sharp drop in clientele. One fairly successful way of analyzing the saturated resin for gold, after drying, pulverising and taking of a more or less homogeneous sample, is to burn off the organic matter and to perform either a cupellation fire assay directly or a crucible fire assay followed by cupellation. You have to show the gold bead from the resin to the bomber immediately. You may note that he will start thinking in more positive terms of scientists in general and of chemists in particular.

More things can be done with the contents of the left hand side tank. The hydrogen peroxide will already have converted most of the lethal cyanide to cyanate, still we will endeavour to destroy all remaining cyanides in the left hand side tank with the addition of simple "swimming pool" chemicals. We can also make certain that the pH of the solution is adjusted to meet environmental affluents standards.

The simple right-left tanks can also be used to recover gold out of spent gold solutions, drag-outs etc - whenever a plating department is available the ion exchange resin technique of recovery can be easily used, perhaps after some pH or other minor adjustments. The bulk of the dried saturated resin can be refined by incineration followed by melting and casting into a bullion bar or button.

A second recovery source is from <u>diamond cutting operations</u> I understand that diamond cut chains reflect the light differently from uncut chains and are therefore more esthetically pleasing to the eye, and consequently more expensive to the purchaser. There are two sorts of diamond cutting machines: conventional ones and "ice machine" diamond cutters. The latter have a specially chemically cooled drum around which the precious metal chains are wound. The diamond does its cutting going continuously from left to right while the drum is rotating at a set high speed. Whatever the cutting mechanics are, the recovery remains basically the same. Minute particles of precious metal are removed from the chains and have to be recuperated prior to refining. The conventional diamond cutters have little section hoods attached which trap the particles as they are flying away. With ice machines however, no flying particles are generated since the cut-off precious metals stick to the rotating drum; they can literally be flushed away from the drum and collected in any reservoir beneath the machine.

It was found that the precious metals chips from both types of diamond cutting operations can easily be combined first. What we have then is a mass of dirty liquid with the majority of the chips lying at the bottom of the receptacle and lots of oil and grease and some more chips floating at the top. The easiest and cheapest solution was to invest in a laboratory type vacuum pump, some suction flasks (or flasks with tubulation as the catalogues would have it). Büchner filters and some appropriately sized analytical type filter paper circles. In reality, a simple filtration under vacuum took place separating the precious metal chips from everything else. Once the chips are collected on the filter paper they can be treated with degreasing chemicals and other cleaning agents if so required. On a daily basis they can, after vacuum filtration, be transferred to stainless steel trays and then dried in a laboratory drying oven. After drying, the finely divided precious metals can be thoroughly mixed. A representative sample can be drawn and the amounts of gold and silver can be established using classical fire assay procedures. As the bulk of the dried material is weighed too, the amounts of precious metal recovered on a daily basis from diamond cutting can easily be tabulated. The material is then ready for refining. I found that, if anything else, the material was too dry at first and encouraged our Vault Personnel to have sneezing fits, which were harmful to my reputation as a precious metal recoverer.

A third and last recovery source I would like to touch on, is the recovery from sludges in <u>deburring and tumbling end tanks</u>. It is definitively the slimiest recovery of all three. All solutions generated by deburring and tumbling are collected in a series of end tanks, each of which has an overflow into an adjacent end tank. After some time greenish slurries settle at the bottom of the first end tank. As time progresses slimes are formed in the other end tanks too. One spectacular day when the slurry in end tank No.1 had risen to a considerable height I was adventurous enough to lie flat on my belly and lower a beaker in the slurry, while the Deburring and Tumbling Supervisor was looking on in utter disbelief. Back in the relative safety of the lab I dried the obscene looking material and then proceeded to pulverise it in a mortar end pestle. Low and behold: even my myopic eyes could clearly distinguish shining particles. After burning off the organic material and carrying out a crucible fire assay fusion, followed by scorification and cupellation I found fairly astonishing amounts of gold and silver.

Getting hold of the bulk of the slurries proved to be a headache. First we had to
pump out the supernatant liquid and then attack the slurries themselves. After a
series of pump breakdowns an employee in hipwaders had to be lowered in the end
tanks armed with a shovel. All slurries were finally collected in drums. I made
a rather foolhardy attack to reduce the volume of the slurries through large scale
boiling, which resulted in my popularity dropping to an all-time low. The slurries
were ultimately incinerated, crushed and sampled and refined as sweeps.

I've only mentioned three sources of recovery in the jewellery manufacturing
industry. Naturally there are a lot more: I could have mentioned sweeps, casting
buttons or donuts and the obvious jewellery scraps. Time wouldn't permit. Neither
would you, fortunately.

Precious metal recovery is one thing, artistry is another and the refining of the
recoveries is yet another thing. All three factors are of enormous importance in
the jewellery industry: they are naturally interwoven and interdependent. The
task of the chemist is to see to it that as much as possible is recovered and to
make certain that all is refined in a proper manner. Maybe the days that jewellery
manufacturing was purely an art are behind us, science is raising its ugly head
now...

A Survey of Precious Metal Conservation Techniques in Electroplating Operations

Joseph J. Werbicki, Jr.
Avon Products, Inc., Mansfield, Mass.

ABSTRACT

For years the users of precious metals have been exposed to conservation techniques. Enlightened users have always practiced conservation. However, it is only when metal costs skyrocket that these techniques are suddenly "discovered" and finally put to use by the majority of users.

The ways 'n which savings can be realized are myriad, ranging from alternative materials to more effective reclamation of scrap.

This paper will touch briefly on a wide variety of precious metal conservation approaches, both novel and time-worn. The reader must objectively evaluate their value as it applies to each individual operation.

KEYWORDS

Conservation, plating, precious metals, substitution, coatings.

INTRODUCTION

For years, platers have been exposed to precious metal conservation techniques. When metal costs skyrocket, as in the past year, basic conservation techniques are rediscovered.

Gold prices were very much on people's minds in early 1980, with London prices hovering at $500 per troy ounce. This reflected a ten-fold increase over the price of gold in early 1972 when gold first broke the $50 mark. A more normalized figure is obtained by comparing the average daily gold price for an entire year, with 1978 averaging $181 per troy ounce and 1979 averaging $300 per troy ounce. The tremendous rise in gold prices at the end of 1979 is, therefore, not fully reflected in the yearly average. Other metals, both precious and non-precious, also fluctuated in price during this time period, with silver the most prominent.

It must be recognized that every conservation technique is not equally applicable to every user. Cost and performance factors are unavoidably interrelated. Therefore, this paper will touch briefly on a wide variety of techniques, and leave the final decision as to their relative values to the consumer.

117

Joseph J. Werbicki, Jr.

CONSERVATION TECHNIQUES

At the outset of a production cycle, <u>part design</u> must be analyzed by a design engineer with an eye toward possible precious metal conservation. Electronic hardware, such as connectors, relays, reed switches, printed circuit board tabs, lead frames, and other components lend themselves to selective plating techniques, such as depth plating, spot plating, and stripe plating.

Other items can be analyzed at the design stage to determine; whether overall gold thickness can be reduced; whether localized thickness standards are more important than overall thickness; or whether substitute coatings can be considered. Perhaps the point most worthy of constant repitition is that all methods to minimize the amount of precious metal must be considered in light of <u>maintaining both performance and quality</u>.

<u>A realistic set of quality</u> standards must be established to take advantage of conservation techniques. United Kingdom printed circuit and connector specifications are reputed to be tighter than those in the United States. The use of low karat golds and other substitute materials is restricted in the electronics industry, while the decorative industry is often slow to adopt new techniques. Simply stated, overspecification can be quite costly.

More attention must be paid to <u>measurement and control</u> practices. Better in-process thickness and plating bath control are needed. Cross-sectioning is still the standard thickness measuring technique[1], given the density variations in electrodeposits, especially with low karat gold alloys. Composite low karat/high karat coatings are difficult to measure individually by any other technique.

Instrumental and chemical methods, such as beta backscatter, eddy current, coulometric and stripping are rapid, but results are influenced by deposit density, surface topography, and instrument calibration. Beta backscatter has the advantage of being capable of measuring extremely thin deposits by the use of newly developed low power isotopes and improved geiger tubes.

A comprehensive study of measurement errors (1) recommended mass standards (mg./sq.in.) as a substitute for thickness standards to compensate for density variations. The National Bureau of Standards (U.S.) thickness standards are sold on the basis of mass per unit area. Matching of standards to the measured coatings and cross-checking non-destructive testing data by stripping and cross-sectioning can produce more accurate measurements.

The technique of matching standards also applies to analysis of plating baths. Rather than analyzing solutions against pure standards, it is advisable to analyze a bath against known-concentration operating bath samples, wherein matrix and interference effects are factored out of the analysis. This is especially important with atomic absorption spectrophotometry.

<u>Improved plating practices</u> are the best approach to sound plating. It is best to specify systems which have the greatest tolerance for abuse before they begin to yield defective deposits. Abuses are generally encountered as current density variations, bath contamination, improper replenishment, poor housekeeping, and poor bath maintenance. Costs can be reduced by using easy-to-control plating baths to minimize both downtime and plating rejects.

<u>Productivity improvements</u> can be a source of savings easily overlooked. While not actually reducing metal usage, minimizing downtime on baths through careful analysis, maintenance and control plays double returns since production time is not lost and defective parts are not produced. Using baths with faster deposition rates and maximizing the parts per rack or barrel gets more work out the door per unit time, provided quality practices are enforced.

---

[1] ASTM B-488 does not recommend cross-sectioning for deposits below 100 micro-inches.

Another area of savings is <u>improved plating distribution</u> on electroplated parts.[2] It is the unavoidable nature of an electroplating process that deposit thickness varies with potential (voltage). Potential varies over a part's surface and part-to-part, due to the combined effects of anode/cathode spacing, part-to-part spacing, part orientation, solution agitation, and other dynamic factors within a plating bath.

Simulation studies have been performed (2, 3) to aid in the design of parts to improve plating thickness uniformity. Other reports have covered anode and cathode masking, tank geometry, thieving, auxiliary anodes, and bipolar anodes (4, 5) which are used in many cases to improve plating distribution. The use of plating baths whose current effeciency decreases with increasing current density also aids in uniformizing deposit thickness.

The overall goal is to reduce overplating to achieve the lowest average thickness which will still meet minimum thickness requirements. The average thickness can be lowered if thickness variation is reduced. By controlling the primary current density distribution (that controlled solely by geometry and spacing), and the use of the above design features, improvements can be obtained. However, work must be done experimentally, as mathmatical models are too complex to be of practical value (6). Plating rack design can be altered to improve current flow to parts, but again on a trial and error approach.

Barrel plating generally produces better thickness distribution that does rack plating, provided solution transfer is optimized and proper bath formulations are used. While barel plating is often reserved for items which are difficult or impossible to plate by any other means, many other items can be barrel plated successfully following a little experimentation (7, 8, 9, 10). In addition to improved distribution, processing savings will also be realized.

The use of <u>pulse plating</u> has been shown to prevent overplating ("dog-boning") in barrel plating of connectors (11), to improve plating in blind holes when combined with ultrasonics (12), and to provide more uniform thickness distribution.

By <u>altering bath chemistry</u> through the use of additives and pH control, the throwing power of a bath can be improved (13). Throwing power is the ability of a bath to plate at a higher efficiency in the low current density (recessed) areas than at the higher current density (exposed) areas, thereby producing a lower thickness variation than would be predicted from geometry alone. Faster-brightening nickel processes also improve productivity, while consuming less nickel.

An A.E.S. research program (14) is presently studying the properties of deposits as they are affected by bath impurities and additives. Other such projects will be cited later.

Durability of plated coatings is affected by (among other factors generally grouped under the term of "Toughness") <u>porosity</u> of the coatings. Often wear-resistant coatings fail prematurely due to corrosion, which can occur through pores in the outer protective coating. Porosity is influenced by base metal surface condition, bath composition, and the presence of suspended matter in the bath (15). The effects of these decrease as the deposit thickness increases. Increasing the deposit thickness to reduce porosity is both uneconomical and impractical. One recent study (16) showed that all gold deposits less than 1 microm thick are porous, while an early study (17) showed that 10 microns of cyanide gold over silver was required to prevent silver tarnish.

Acid golds are inherently less porous than cyanide golds (16). At heavier deposit thicknesses ($\sim$100 μin) sulfite (non-cyanide) gold deposits are less porous than acid gold deposits. However, the wear resistance of sulfite gold deposits is poor. Acid gold has superior wear which may be due to surface chemistry effects and intrinsic lubricity due to co-deposited polymers.

[2] The term "electroplated" is used because the use of elecroless (autocatalytic) plating (where justified) does produce uniform deposits.

Porosity testing is a valuable tool, in conjunction with life-time wear testing for evaluating coatings. Chemical tests (18) and electrographic tests (19) are employed to generate comparison data.

Another method for increasing corrosion resistance is the strategic use of <u>preplates</u>. Aside from electropolishing a base metal, which has limited practical application, and keeping in mind the potential problems that can occur in abrasive finishing due to embedding of abrasive grains (16, 20), the simplest way for a plater to reduce porosity is to improve the surface of the base metal by the use of levelling base plates. Acid copper is known for its exceptional levelling, while bronze has been found to improve corrosion resistance, due partly to levelling and partly to reducing galvanic corrosion (21).

Savings can be realized by using plating baths with <u>lowered metal contents</u>. Gains to be realized here are; reduced tank make-up and inventory costs; lower metal dragout losses; and decreased waste treatment and recovery costs. Specific examples are acid gold baths which operate at 0.25 troy ounces per gallon, trivalent chromium, and low-temperature nickels.

Some savings can be realized by looking to bath suppliers for <u>improvements in deposit technology</u>. Formulae with improved throwing power, covering power, and levelling, produce <u>deposits</u> that achieve, in thinner deposits, performance equal to thicker coatings. To counteract the effects of porosity, gold bath studies on deposit orientation (22) have shown how to control deposit growth to improve covering power.

Partial and complete <u>substitution</u> for precious metals is currently under widespread considera-tion. Low karat <u>gold deposits</u> (8 karat to 18 karat) have very impressive cost savings associated with their use as substitutes for 22-24 karat gold deposits. Copper-cadmium golds are popular in Europe with watch case manufacturers. However, the toxicity of cadmium may militate against its use in the U.S., where silver-gold alloys are popular. Wear resistance of these alloys is not as good as that of hard acid gold deposits and the deposits are susceptible to discoloration if uncoated. Both of these drawbacks are overcome by topcoating the alloy gold with a high (22-24 kt) karat hard acid gold. While a topcoat of 20 microinches is recommended for sulfide environments (23), even lower thicknesses have been found to be adequate in many applications.

Wear resistance increases as the acid gold topcoat thickness increases, so it is important to balance cost reduction against desired wear properties. Unless savings are greater than the wear ratio in a specific application, there may not be a net gain realized.

Palladium/nickel alloys (50/50 and 80/20) are being substituted for rhodium and gold. The alloys provide excellent contact and connector surfaces, have high plating rates, good corrosion resistance, high hardness, and are as good as gold in some applications. Palladium/ nickel is even used to replace nickel under gold and rhodium to produce better overall corrosion resistance.

Palladium is replacing rhodium, either entirely or partially, when followed by a thin top coat of rhodium in decorative applications. Palladium has good wear resistance, solders well, is a diffusion barrier (24), and is relatively inexpensive when compared to other precious metals. Older bath formulations produced a rather dark deposit, but recently-developed baths are bright and quite white.

Ruthenium is relatively inexpensive and readily available. It has excellent wear resistance, and forms a conductive oxide. Ruthenium/rhodium alloy deposits are used on watch bezels to match the color of stainless steel watch bands.

There are applications where non-precious metal coatings are used as precious metal substitutes, i.e. tin and tin/lead coatings provide functional coatings in some electronics applications replacing gold; brass and bronze deposits are used to simulate gold on many items (these are often clear coated for tarnish resistance); and so-called artificial rhodiums (bright tin and speculum bronzes) replace rhodium on inexpensive decorative items.

Protective coatings can be used in conjunction with lowered precious metal thickness or with precious metal substitutes, especially in decorative applications where solderability at a later date is unimportant. The best of the clear coatings require baking to attain maximum properties of corrosion and wear resistance. However, O.S.H.A. regulations on emmisions and safety add costs to the use of solvent-based formulations. Alternative water-based coatings may overcome these drawbacks, while providing lesser, but adequate protection.

Whenever conservation is discussed, the subject of accountability must be addressed. A recent short article (25) enumerated losses of gold due to deposits on non-functional parts, lack of security, analytical errors, disposition of lab inspection samples and thickness testing errors.

Not only is it advantageous and desirable to use less precious metal or to use substitutes, but more effective reclamation and accountability techniques can be a source of major savings.

Reclaim and reuse of plating solutions are effected by a variety of techniques, including dragout tanks, ion exchange, evaporative recovery, reverse osmosis, electrodialysis, and electrolytic recovery. Removal of metals from the effluent stream can also contribute to a reduction in waste treatment costs.

Other techniques that can save money are; stripping of rejects, especially those to be replated; reducing plate on pins, wires, danglers, and other non-work surfaces; dragout reduction techniques (26); stripping of pins, danglers, etc. in-house; returning dragout to plating tanks to save market, fabrication and processing costs vs. only market costs when refining dragout solutions; recovery of gold and rhodium from carbon and filters. Finally, don't forget base metal values in refining lots, tin for example.

In-process savings should be credited to the plating department, otherwise personnel might be inclined to settle for end-of-process recovery to document their efforts in the area of reclamation.

Scrap recovery should be co-ordinated by a single individual who is aware of the many areas of precious metal scrap generation and who is knowledgeable in the various recovery techniques that can be used to extract precious metal values at the most economical rate.

While precious metal scrap recovery programs are quite straightforward and can be implemented with the assistance of reputable refiners, the amount of precious metal that is deposited on production parts must be closely monitored. Some plating baths change their deposition rate as pH changes occur. Unless such solution parameters are monitored quite effectively and corrections made, the amount of deposit will not be that which is programmed for an established plating cycle.

Continuous in-process checks must be made to verify that all established parameters are working to produce the desired results. Timely analytical feedback to production is mandatory.

An end-of-pipe check on precious metal usage involves the use of costing data for production items. The weight of precious metal, as costed for a piece, can be entered into a computer. As pieces are produced, the quantities should also be entered. For any given period of time, the computer can produce a projected total precious metal usage, which can be compared to the actual amount drawn from inventory. Costing should include a dragout loss figure. Recovery from scap reclamation can be entered as a credit against dragout in the overall cost analysis.

It must be understood that production processes cannot be changed overnight. There is a learning curve associated with any decision to make a major change in a plating operation. The safest approach is to change over in stages if there is no proven approach available. However, when the information is available from dependable sources, it is also a major cost-saving to take advantage of that proven data.

Precious metal conservation techniques are straightforward. However, all the alternatives should be considered, and a common sense plan developed which will balance the pluses and minuses in order to produce overall savings while maintaining quality.

REFERENCES

1. Rosenburg, S.R., and R.L. Cooley (1970). Quality Control of Precious Metals Plating. Metal Finishing , 68, No.4, 34-38.
2. Kinney, G.F., and J.V. Festa (1954). Current Density Distribution in Elecroplating by Use of Models. Plating, 41, No. 4, 380-384.
3. Christie, J.J. and J.D. Thomas (1965). Plating, 52, No.9, 855-859.
4. Mohler, J.B., The Geometry of Plating Baths (1976). Metal Finishing, 74, No. 11, 53-56; 74, No. 12, 40-42. (1977). Metal Finishing, 75, No. 1, 40-41; 75, No. 2, 80-81.
5. American Electroplaters' Society Illustrated Lecture, Design Precepts for Quality Plating.
6. Kasper, C. (1940). Trans Electrochem. Soc., 77, 353; 78, 131. (1942). Trans Electrochem. Soc., 82, 153.
7. Craig, S.E., Jr., R.F. Harr, and P. Mathieson. The Theory of Metal Distribution During Barrel Plating. Research Report No. 34, Americal Electroplaters' Society.
8. Nobel, F.I., and co-workers (1966). Factors that Influence Gold Metal Distribution in Barrel Plating. Plating, 53, No. 9, 1099-1104.
9. Nobel, F.I., and D.W. Thomson (1970). Metal Distribution in Barrel Plating: Further Studies. Plating, 57, No. 5, 469-474.
10. Nobel, F.I., R.B. Kessler and D.W. Thomson (1971). The Use of Statistical Methods to Reduce Costs in the Barrel Gold Plating of "Difficult to Plate" Electronic Components, Plating, 58, No. 12, 1198-1202.
11. Greig, W.J., R. Brown and W.E. Berner (1976). High Density Multi-Chip Hybrid Microcircuits Using Electroplating Techniques. AES Proceedings of the Design and Finishing of Printed Wiring and Hybrid Circuit Symposium.
12. Duva, R. (1979). Pulse Plating: The Pros and Cons of Practical Applications. Finishing Highlights, Jan/Feb.
13. Rothschild, B.F. (1979). Factors Involved in the Development of Plating Solutions with High Throwing Power. Plating & Surface Finishing, 66, No. 5, 70-73.
14. Weil, R. (1979). American Electroplaters' Society. Research Project 38A, Some Property Structure Relationships in Electrodeposits.
15. American Electroplaters' Society. Research Project No. 6.
16. Page, J.K.R. and V.G. Rivlin (1980). Stable Gold Alloy Electrodeposits, Gold Bulletin, 13, No. 1, 43-48.
17. Harding, W.B. (1960). The Tarnish Resistance of Gold Plating over Silver. Plating, 47, No. 10, 1141-1145.
18. Nobel, F.I., B.D. Ostrow and D.W. Thomson (1965). Porosity Testing of Gold Deposits, Plating, 52, No. 10, 1001-1008.
19. Noonan, H.J. (1966). Electrographic Determination of Porosity Testing in Gold Electrodeposits, Plating, 53, No. 4, 461-470.
20. Gillespie, L.K., and F. Clay (1975). Deburring Effects on Plating Adhesion. Finishing Highlights, Sept./Oct., 26-27.
21. BBX-2 Technical Report for Industrial Applications (1979). American Chemical & Refining Co.
22. Morrissey, R.J., A.M. Weisberg and H. Shoushanian (1976). Method of Reducing Porosity in Gold Deposits, AES Symposium-Design and Finishing of Printed Wiring and Hybrid Circuits.
23. Shoushanian, H. and A.M. Weisberg, Low Karat Silver-Gold Alloy Deposits, Unpublished Report.
24. Pickering, H.W., W.R. Bitler and D.R. Marx (1977). American Electroplaters' Society. Research Report No. 29.
25. Donaldson, J.G. (1980) You'll Wonder Where the Yellow Went. Plating & Surface Finishing, 67, No. 3, 16.
26. Kushner, J.B. (1970). Water and Waste Control for the Plating Shop, Gardner Publications, Cincinatti, 50-56.

Analysis

Inductively Coupled Plasma (ICP) in a Precious Metal Laboratory

B. Horner.

Johnson Matthey Chemicals, Enfield, Middx., England.

ABSTRACT

The introduction of a computer controlled inductively coupled plasma spectrometer into a precious metal laboratory is described. Comparisons are made with existing analytical methods where AAS is used for final measurement. Several benefits arise from using a simultaneous emission technique. The extended dynamic range made available avoids the need for serial dilutions. Also the freedom from chemical interference in the source allows measurements to be made without recourse to releasing agents. Sensitivities for all precious metals are as good as or better than those by flame AAS and no significant spectral interference has been found between precious metal elements. Some interferences have been encountered from spectral overlaps of other elements present in solution and a means to correct for these is outlined. The general application of ICP to the assaying of residue materials is covered.

KEYWORDS

Inductively coupled plasma, precious metals.

INTRODUCTION

The role that Johnson Matthey has played in the field of precious metals refining has been and still is considerable. However the UK involvement in this area is a changing one with in recent years an increased volume of secondary refining of precious metals particularly gold and silver replacing the extractive mining processes and primary refining much of which is now done at source in South Africa.

Consequently the analytical requirements have needed to expand to embrace these new interests. The role of the laboratory in this time has changed from one of largely process control and the analysis of well defined plant intermediates to one where there is a much greater involvement within the laboratory on the evaluation analysis of incoming customer material.

This may take the form of photographic waste, electronic scrap, jewellers sweeps, refinery residues, spent catalyst or in general any precious metal bearing material suitable for processing that can be reduced either by burning, drying, grinding and blending to a homogeneous powder form suitable for analysis.

125

Material may also be received in a metallic form or metallic products generated
from sampling processes and these induction melted to produce a sample for assay.

Where materials cannot be melted or conveniently sampled, an acid digestion may
be effective and produce a solution sample for analysis. Solutions, largely from
plating processes, may also be received directly from the customer.

In the subsequent processing of residue materials via a smelting circuit as the
first stage of recovering the values present, it is highly desirable to know the
fluxing qualities of the material to be smelted. If information on the refractory elements is to hand a balanced furnace charge can be composed from the
residues available. This in turn reduces the amount of additional fluxes to be
added and minimises losses into slags, thereby increasing the blast furnace
efficiency.

With the general increase in information requested and the need for a quicker
outturn on all evaluations it was decided to purchase an inductively coupled
plasma direct reading spectrometer (ICP). In the previous ten years many of the
analytical procedures employed have been tailored to produce a total solution
for measurement by atomic absorption. The advent of AAS was a great benefit to
all those involved in the precious metal field, but where the volume of work was
increasing and where multi-element determinations were requested it suffered
somewhat from being a sequential technique. It had been this laboratory's
experience that with three AA sets all with turntable feed and with data collection and processing carried out by a small dedicated computer the department was
always working at full capacity. As nearly all existing preparation techniques
could be retained, a ready comparison between figures obtained by AAS could be
made with those obtained by ICP. This was a great benefit as it allowed results
from ICP to be accepted and used ahead of what would be normally expected when
setting up a new instrumental technique. Thus it was the initial target to at
least match figures obtained by an accepted technique and in time to use this as
a base to improve on both their precision and accuracy.

GENERAL CONSIDERATIONS

It is convenient at this point to catalogue some of the inherent advantages of
ICP over AAS and also to mention some aspects which might be thought disadvanta-
geous. While these observations do have a theoretical basis they are given simply
as points for consideration by any potential user.

The over-riding benefit comes from simultaneous measurement of as many elements
as are catered for by the spectrometer. Being an emission technique the linear
dynamic range is generally better than 10^4 and this allows bulk solutions to be
aspirated directly without prior dilution as required for AAS. This represents
a considerable saving in time. Also whereas calibrations with AAS often needs
second and third order regression most ICP calibrations can be fitted to a
straight line.

The greater source temperature means that chemical interferences in the plasma
are rarely if ever encountered. In AAS especially with precious metal solutions,
releasing agents or ionisation buffers are often required.

The use of an inert plasma and carrier gas means that there is no reducing
atmosphere in the spray chamber therefore deposits and potential contaminants do
not build up. This is a particular problem with easily reducible silver and gold,
a point rarely mentioned in publications.

Limits of detection have been determined and are given in Table 1. Comparisons with those achieved by AAS are given in Table 2.

TABLE 1   Limits of Detection for E1000 Spectrometer-ICP Source

Element	Wavelength (nm)	LOD (mg/l)	Element	Wavelength (nm)	LOD (mg/l)
Ag	338.3	0.028	Mo	281.6	0.013
Al	396.1	0.024	Na	330.3	6.7
As	193.7	0.43	Ni	231.6	0.044
Au	267.6	0.008	Os	290.9	0.008
B	249.8	0.019	Pb	220.4	0.14
Ba	455.4	0.002	Pd	340.4	0.014
Be	313.0	0.011	Pt	265.9	0.044
Bi	306.8	0.16	Rh	339.6	0.010
Ca	317.9	0.011	Ru	349.9	0.047
Cd	226.5	0.009	Sb	231.1	0.12
Cr	359.3	0.018	Se	196.0	0.62
Co	228.6	0.030	Si	288.1	0.053
Cu	327.4	0.089	Sn	284.0	0.087
Fe	259.9	0.012	Te	238.6	0.36
Hg	253.7	0.048	Ti	334.9	0.004
Ir	263.9	0.096	W	276.4	0.13
Mg	280.3	0.003	Zn	213.8	0.020
Mn	257.6	0.004	Zr	343.8	0.010

LOD Expressed as 2 standard deviations at zero concentration.

TABLE 2   Comparative Limits of Detection ICP vs AAS

Element	ICP[1] (mg/l)	AAS[2] (mg/l)
Au	0.008	0.010
Ag	0.028	0.003
Pt	0.044	0.12
Pd	0.014	0.03
Ir	0.096	0.36
Ru	0.047	0.16
Rh	0.010	0.010
Os	0.008	0.12

1.   Rank Hilger E1000 Spectrometer - Plasma Therm Source.
2.   Varian AA6 Spectrometer (Flame AAS)

The factors that most deter potential users are those concerned with the interference problems that can arise from using an emission technique. Many of these can be minimised by the correct choice of line and by using a spectrometer capable of high resolution. ICP being an emission technique does not have the specificity of AAS. This in practice means that information on the other elements present in solution is required. Because of the simultaneous measurement of these other elements this otherwise redundant information is often available and can be used to make these corrections. This will be mentioned further later.

The purchase of an ICP direct reading spectrometer represents a considerable investment for any laboratory and any such decision will depend on the volume of multi-element analysis that is required. A considerable financial saving is made if a plasma source can be fitted to an existing spectrometer although the lines available may not be those best suited to ICP. The running costs are largely those of argon gas comsumption and in this laboratory work out to be £1-2 per hour.

INSTRUMENTATION

The spectrometer is a Rank Hilger model E1000. This is a twin holographic grating vacuum path instrument with the longer wavelength dispersion effected by a periscope optical system. Both gratings have a reciprocal dispersion of less than 0.3 nm/mm in the first order and between them cover the wavelength range 190 - 450 nm. Forty five measurement channels are employed covering thirty seven different elements. The 45th channel monitors the output from a monochromator set at right angles to the main optical path. This allows elements not on the main analytical program or those beyond the range of the gratings to be included. The monochromator also allows chosen lines to be scanned in order to detect possible spectral interferences. The dispersive power of the monochromator is only one third that of the main gratings.

A standard Plasm Therm 2.5kw RF generator operating at 27.12 MHz powers the plasma torch. The torch assembly has been remodelled to fit in with the optical lay-out of the spectrometer and uses argon as both coolant and plasma gas. A forty position sample changer is used to feed samples and wash water alternatively. A peristaltic pump is used to force feed a Meinhard concentric glass nebulizer. The nebulizer gas is humidified by passing the argon gas stream through water. These are measures taken to enable a continuous feed without the salt content of the solution blocking the capillary of the glass nebulizer. Under such conditions it is possible to introduce samples containing up to 2% solids without a dramatic fall in nebulizer efficiency.

The position of the plasma in relation to the entrance slit of the spectrometer can be varied to allow parts of the plasma to be examined. A viewing height 10mm above the coil is found by measuring intensities for different elements at varying heights to be an effective compromise on most analytical programs. Unfortunately the operating parameters for an ICP are inter-related and a similar exercise is necessary for the incident power, coolant, auxiliary and nebulizer gas flow rates for each element. This means an inconveniently large number of measurements must be made if all combinations are to be investigated. However, patterns between elements do appear and it is possible to choose compromise conditions on all parameter settings.

Typical operating conditions:

Incident power	1.6 kW
Auxiliary argon	0.5 l/min
Nebulizer argon	0.5 l/min
Coolant argon	10.0 l/min
Observation height	10mm
Solution uptake	2.5 ml/min

The sequence of measurement is controlled by a small computer accessed from a VDU terminal. The required analytical program is called up and the spectrometer conditions selected. This identifies the channels to be read, those to be used for internal referencing, delay and integration times together with EHT settings for each selected channel. On completion of the run sample concentrations are

calculated from intensity measurements and previously entered standard data.
This is matched with additional sample information concerning weights and
volumes to produce a final printed report. The read cycle normally employed
incorporates a 15 second delay time plus a 15 second integration time. Between
four to six standards are used to cover the range of sample concentrations
expected. Several runs may be compared and accumulated and the system allows
inspection of intensity data at any time.

The computer (Trivector Systems) also independently collects and processes data
 from an XRF spectrometer, a potentiometric titrator and two analytical balances.

PROCEDURES

As previously mentioned, traditional methods of preconcentration are used in order
to obtain a clean solution for measurement. The choice of method depends on the
elements required and the nature of the material to be assayed. It is necessary
to include these separations on evaluation analysis in order to achieve the
accuracy required. An outline of these procedures is given.

For low grade residues where rhodium, ruthenium and iridium determinations are
not required a lead collection by fire assay is used. Cupellation followed by
a nitric acid parting subsequently leads to a solution containing gold, platinum
and palladium suitable for ICP requirement.

Where residues contain rhodium, ruthenium or iridium a nickel sulphide collection
is preferred. After parting in a hydrochloric acid/ammonium chloride mixture
the residue can be dissolved in aqua regia or fused with sodium peroxide to
obtain a total solution for analysis.

For higher grade residues a direct peroxide fusion is followed by distillation
to remove ruthenium and osmium. The remainder values are recovered from
solution firstly by hydrolysis and then by zinc reduction. The combined recoveries are re-dissolved to obtain a solution suitable for ICP measurement.

The reproducibility of a cross section of elements in aqueous solution is given
at two levels in Table 3. Where chemical procedures have preceeded the final
measurement it is found that more reproducible results are obtained if intensities are internally referenced before calculation. This is achieved by adding
an indium solution to both samples and standards to give a final concentration
of 100 ug/ml. The use of an external standard compensates for small fluctuations
in acid strength and nebulizer efficiency.

TABLE 3   Reproducibility Data for ICP

Al	1 ± 0.006	5 ± 0.03
Au	1 ± 0.01	5 ± 0.04
Ca	1 ± 0.01	5 ± 0.04
Cr	1 ± 0.01	5 ± 0.04
Cu	1 ± 0.01	5 ± 0.03
Ir	1 ± 0.03	5 ± 0.05
Mg	1 ± 0.02	5 ± 0.05
Mo	1 ± 0.01	5 ± 0.04
Pd	1 ± 0.005	5 ± 0.03
Ti	1 ± 0.01	5 ± 0.04

All figures are in mg/l $\pm$ one standard deviation.

INTERFERENCES

It has been necessary to establish on the methods examined so far that other elements present in solution do not produce an interference. Information on background equivalent concentrations of spectral interferences on the precious metals have been compiled and these are to be incorporated in the analytical programs employed. The information required to apply these corrections is readily available from other channel intensity measurements and these can be used to modify the analyte intensities before concentrations are calculated. A list of known effects greater than 1 part in 1000 is given in Table 4.

TABLE 4 Spectral Interferences on Precious Metals

Background equivalent concentrations (mg/l)

Au(267.7)	Ag(338.3)	Pt(265.9)	Pt(299.8)	Ir(263.9)
Ru 1.6	Ti 1.7	Cr 3.4	Mo 27	Mn 97
Cr 1.5		Ru 1.8	W 2.5	Mo 39
W 1.5		Mo 1.5	Cr 1.2	Zr 13
			Cd 1.2	Fe 4.4
			Mn 1.0	Ru 3.6
				Cd 2.4
				Os 1.7
				Cr 1.5

Pd(340.4)	Ru(349.9)	Rh(339.6)	Os(290.9)
Zr 3.0	Zr 2.4	Mo 4.1	Mo 40
	Rh 2.0	Zr 1.4	Ag 2.2
	Mo 1.6		Cr 1.2
	Ti 1.2		

Each interferent measured at 1000 mg.l

In summary, the equipment, which has been in use for six months has proved reliable and effective and I am sure could be an aid to accomplish multi-element determinations in other laboratories as it has been in this.

Developments in Iridium Analyses

A. Manson and St.J. H. Blakeley

Inco Metals Company
J. Roy Gordon Research Laboratory
Sheridan Park, Ontario, Canada

ABSTRACT

Developments in the determination of iridium over the last decade are discussed reflecting the broadening of the range and types of materials which can now be handled. Concentrate analysis time is reduced from about two months to several hours by means of atomic absorption and most process solutions can be assayed in about ten minutes. The analysis of low grade solids by emission spectroscopic analysis of platinum beads produced by fire assay techniques is described.

KEYWORDS

Analysis; iridium; precious metals; atomic absorption; emission spectroscopy.

INTRODUCTION

The ores of the Sudbury Basin have been mined for many years producing not only copper, nickel, cobalt, selenium and tellurium but significant quantities of platinum group metals, gold and silver. These precious metals are initially present in extremely low concentrations but become more concentrated during processing. Eventually these valuable components are isolated into a by-product concentrate suitable for further treatment and refining. Until recently, the complexity of analytical procedures denied the assaying of ores and most of the intermediate process materials for iridium and only the high grade concentrates were analysed on a regular basis for accounting purposes.

Of the platinum group metals iridium is probably the most difficult to determine. This paper reviews some of the developments over the last decade.

CLASSICAL METHOD

The old classical method for the determination of iridium is a long, involved procedure requiring many separations. On materials such as precious metals concentrates, the steps involved are smelting, lead collections, slag cleaning, parting, silver removal, lead removal, nitrite complexations, bromate hydrolysis, selective precipitation of platinum, palladium, gold and rhodium followed by separation of ruthenium by distillation before the final gravimetric determination of iridium. The time for analysis by such a procedure was usually about eight weeks. Consequently, most low grade process intermediates were not analyzed for iridium at all, and amongst the

131

higher grade products usually only analysis of final concentrates was done on a re-gular basis.

Generally, iridium is a minor component in comparison to base metals, platinum, palladium and even rhodium. Thus separation steps must be taken before colorimetric methods can be applied. Such separation procedures have included solvent extraction and ion exchange but all are complex and lengthy.

The traditional fire assay collection of precious metals into a silver bead for spectrographic analysis is suitable only for platinum, palladium and gold. Even collection into a lead bead is not satisfactory for iridium due to inhomogeneity.

The advent of atomic absorption offered a more simple and rapid method of analysis applicable to a wide range of materials.

ATOMIC ABSORPTION ANALYSIS

We first used atomic absorption for the analysis of high grade materials in 1965 to provide rapid palladium, platinum and gold data for a particular research program. During the next few years improvements in atomic absorption instrumentation and practices gave greater precision. Between 1969 and 1971 there was increased interest in precious metals analysis and a routine method for the determination of platinum, palladium, rhodium, ruthenium, gold and silver in concentrates was established. The analysis time was reduced from eight weeks to about three hours and agreement with classical results was very good.

Analysis of Concentrates

The procedure devised, with speed, accuracy and simplicity in mind, consisted of fusion of 0.500 g of sample with 5.0 g of sodium peroxide to form a melt which upon water leaching and acidification with hydrochloric acid gave complete dissolution. After making up to 500 ml volume, aliquots were diluted to give solutions containing approximately 50 mg/l Pt and Pd, 10 mg/l Au and 7 mg/l Rh and Ru and also 5 g/l Cu as sulfate. Copper was omitted from solutions for palladium analysis as it is not a necessary additive for this element and can in fact be detrimental particularly in the presence of silica. These were read against standard solutions of a similar matrix and containing precious metals bracketing the concentration in question. The result was a rapid, precise and accurate method.

DETERMINATION OF IRIDIUM BY ATOMIC ABSORPTION

Iridium was not included in the primary method described above as the sensitivity was poor and the instrument noise levels were high. The relative content of plati-num group metals and their relative sensitivities are given in Table 1. The poor sensitivity for iridium together with the relatively low concentration immediately indicated that some modification to the general method was necessary.

TABLE 1 Relative Concentrations and AA Sensitivities of
        Precious Metals

Element	Ir	Pt	Pd	Rh	Ru	Au	Ag
Rel. Concentration	1	17	17	2.5	2	3	1.5
Rel. Sensitivity*	240	50	4	10	10	10	1

*Concentration required to give equal absorbance readings

A general investigation of the effect of base metals and other platinum group metals on iridium showed no interferences provided that copper sulfate and sodium chloride are added in the same proportions as for the analysis of other precious metals, i.e. 5 g/l Cu and 5 g/l NaCl. Higher concentrations of copper did not affect the absorbance.

The absorbance values for iridium are, like ruthenium, greatly influenced by flame conditions and it was found that careful optimisation of the acetylene to give a slightly luminous flame was necessary.

Three possible procedures were tried,

a) Direct reading of the solution prepared for platinum analysis using scale expansion at 264 nm.

b) Pre-concentration of iridium by precipitation using bromate hydrolysis, followed by redissolution in acid; no scale expansion required.

c) Dissolution of the bulk of the material by digestion in aqua regia and fusion of the small amount of insolubles.

Procedure (a)

Fairly good results were obtained using the scale expansion technique. Four standard solutions, containing 10, 12.5, 15 and 17.5 mg/l Ir, 5.0 g/l Cu as sulfate, and 5.0 g/l NaCl, were prepared for the atomic absorption measurements. Using the concentration mode the 10 mg/l standard value was expanded so that the read out on the Perkin Elmer 403 gave 10.0 units. Each of the samples was then read using the 100 average mode, immediately followed by the closest standard solution. The value obtained for the standard solution was then used to calculate the unknown value. This method gave results which when compared with results from x-ray fluorescence and classical analysis showed a tendency to be biased high, as shown in Table 2.

Procedure (b)

This procedure enabled us to work at a higher iridium concentration, thereby avoiding scale expansion, by employing a hydrolytic technique to precipitate iridium together with palladium, rhodium and other minor constituents. The solution obtained by fusion and acidification of 0.500 g of concentrate was treated with sodium bromate and sodium bicarbonate to a pH of 8.0 and the precipitated solids recovered in a glass filter crucible. The precipitate was redissolved in hydrochloric acid, made to a volume of 100 ml and measured in the normal manner. Generally acceptable results were obtained though the method was somewhat lengthy and cumbersome.

Procedure (c)

The third method proved to be superior to those previously described enabling a larger weight of sample to be used again avoiding scale expansion and higher noise levels. A 5.00 g sample was leached in aqua regia for about one hour and the insoluble residue filtered off and washed well with hot distilled water. The filter paper and solids were dried in a zirconium crucible and then ignited to burn off the paper. The residue was intimately mixed with 5.0 g of sodium peroxide and heated over a burner until a cherry red clear melt was obtained. The aqua regia filtrate meanwhile was evaporated to a very small volume and chloridised. The fusion product was cooled, leached with water, acidified with 40 ml of hydrochloric acid, boiled and combined with the chloridised aqua regia leach solution in a 500 ml volumetric flask containing 2.5 g of copper (as copper sulfate) and then made up to volume.

Standard iridium solutions of 100, 125 and 150 mg/l Ir, and containing the requisite

copper sulfate and sodium chloride, were made to read 1.00, 1.25 and 1.50 respectively
by using the concentration mode and expansion. The samples were then aspirated and
the iridium values read directly in percent. The results obtained by the three
methods are compared with each other and with those by classical and XRF methods in
Table 2.

The obvious advantages of method (c) are: (i) no separations are involved thereby
reducing the possibility of losses, such as may occur with the bromate hydrolysis
method, (ii) an improved precision of measurement since no great scale expansion is
used, (iii) the solution is suitable for the analysis of the other precious metals
by appropriate dilution.

This method was successfully employed for nearly four years until it was found that
much greater sensitivity for iridium could be obtained from a Techtron 1200 atomic
absorption spectrophotometer.

TABLE 2 Comparison of AAS Results with Classical
and XRF Determinations

Sample	Classical	XRF	AA(1)	AA(2)	AA(3)
1	1.16	1.26	1.22	1.22	1.28
2	1.40	1.49	1.56	1.43	1.51
3	1.26	1.35	1.36	1.27	1.34
4	1.22	1.22	1.35	1.17	1.21
5	1.34	1.40	1.42	1.36	1.39
6	1.56	1.69	1.60	1.71	1.63

AA(1) Using scale expansion
AA(2) Bromate hydrolysis
AA(3) Aqua regia leach followed by fusion of insolubles

Using the 208.9 nm line and the Techtron 1200 as little as 5 ppm Ir in solution
could be accurately measured provided the salt matrix was reasonably well matched.
This naturally opened up a whole new field of analysis enabling many types of solids
and solutions previously considered to be too low for atomic absorption analysis to
be assayed. For precious metal concentrates, 0.500 g samples could be analysed by
direct atomic absorption without scale expansion and with improved precision.

Analysis of High Silver Materials

Unfortunately, from the analytical point of view, changes in processing produced
concentrates that increased in silver content from 1.5% to in excess of 20% together
with a decrease in the overall iridium content which led to a reevaluation of the
procedure. It was deemed necessary to remove silver from the system. The accepted
practice at that time was to leach the filtered, washed, aqua regia residue with
ammonia to dissolve silver as the ammine complex and, using double precipitation,
recover any occluded precious metals values. However, with this technique consistent
ly low values of 0.59% to 0.60% iridium were obtained for a concentrate analysing
0.64% by XRF. Silver chloride can be an efficient collector for small amounts of
iridium and an investigation substantiated this claim as shown in Table 3.

TABLE 3 Precious Metal Losses to AgCl Precipitates

	% Lost					
	Pt	Pd	Au	Rh	Ru	Ir
Single pptn.	4	1	<1	13	4	19
Double pptn.	<1	<1	<1	<1	<1	15

The losses due to co-precipitation were eliminated by preleaching with dilute nitric acid to remove the bulk of the silver, (iridium and calcined iridium oxide are insoluble in this acid), followed by fusion of the residue with sodium peroxide to obtain a final solution containing 20% HCl and 5 g/l Cu. When solutions prepared in this manner were read against matrix matched standards, a reproducible value of 0.64% Ir was obtained. This modification was put into general practice for all high silver content samples. When silver was known or suspected to be present as silver chloride a preliminary hydrogen reduction at about 500°C was done to reduce it and any occluded iridium to metal.

DETERMINATION OF LOW LEVELS OF IRIDIUM

Since we were now determining much lower concentrations of iridium by atomic absorption attention was turned to the analysis of leach liquors and barren solutions. Using solutions containing 25 mg/l Ir, 5 g/l NaCl, 5 g/l Cu (as sulfate) and 5% HCl no significant differences in absorbance values could be measured in the presence of 5.0 g/l Ni as chloride or sulfate, 5.0 g/l Fe as chloride or sulfate, 10 g/l H_2SO_4, 1% HNO_3 or 0.1 g/l SiO_2. From this encouraging data the analysis of solutions was undertaken on a regular and routine basis.

Solutions of high iridium content were no problem. Aliquotting to reasonable salt levels, e.g. to <10 g/l, ensured precise, accurate and rapid analyses provided the measured iridium content was in excess of 10 mg/l. Very low levels of iridium, e.g. less than 20 mg/l in the starting solution can, and usually do if the salt concentration is high, give erroneously high values unless corrective measures are taken. This is demonstrated in Table 4.

TABLE 4 Iridium in High Salt Solutions (mg/l)
(Dilution 15 ml to 25 ml)

Direct AAS	44	33	28	24	19	14	14	13
Known Value	43	29	32	23	15	8.6	8.4	9.2

Direct AAS	12	11	11	10	10	8.5	8.0	6.0
Known Value	8.4	5.8	5.4	6.4	6.4	4.8	3.0	3.0

These numbers show that at the higher levels two methods gave acceptably close results but as the concentration drops the results become further apart due to the background absorbance which is fairly constant. The simplest corrective measure employed was the addition of a known amount of iridium to the sample solutions to raise the absorbance values significantly above the areas affected by background. This technique was also sometimes used for iridium solutions of low salt concentration to ensure that the correct values were being reported.

Table 5 compares some apparent results obtained in the analysis of high salt solutions with the more meaningful results obtained by the addition technique.

TABLE 5 Ir in High Salt Solutions by the Addition Technique
(Dilution 15 ml to 25 ml)

Sample	Calculated Assay, mg/l Ir			
	+0 mg/l Ir	+10 mg/l Ir	+15 mg/l Ir	Pt bead
1	7.7	3.2	3.0	3.0
2	3.3	1.3	1.3	1.4
3	3.3	<.5	<.5	.4
4	3.3	∿.8	∿.8	.7

The need for a method of determining iridium in low grade solids, ores, slags, process intermediates, etc., was satisfied by co-workers in the Copper Cliff Process Technology Department who developed a procedure of optical emission spectroscopic analysis of platinum beads produced from cupellation of lead buttons containing added platinum. In general the standard fire assay practice of fusion with lead containing fluxes, to which has been added 10 mg of platinum, is carried out followed by cupellation for removal of the lead.

EMISSION SPECTROSCOPIC ANALYSIS

Sample beads containing from 0.1 to 25 μg of iridium are arced together with a set of seven standard beads covering the whole range. The spectrograph used is a three metre Baird Atomic concave grating model TX-1 and the emissions are recorded on photographic plates. The general parameters are

Instrumental Parameters

Discharge	DC Arc 15A
Analytical gap	4mm, maintained
Slit width	10μ
Rotating sector	100 and 25% transmittance
Pre-exposure	none
Exposure	50 seconds
Sample electrode	Ultra Carbon 213302
Counter "	.120" graphite rod
Sample polarity	positive
Photographic plate	Kodak SA No. 1

Element	Analytical Line (A)	Sector	Range μg
Iridium	2924.79	100%	0.1 - 3.0
	2924.79	25%	1.- - 10
	2934.64	25%	2. - 25

Background readings are taken for all lines and are measured to the low Angstrom side of the line.

Calculation Procedure

The optical transmittance readings obtained from the photographic plate are converted to log intensity ratios using an Artisto Respectra calculating board. The background intensity ratio is subtracted to give a net value. Calibration curves of intensity ratio versus concentration are drawn using data obtained from the standard platinum beads. Iridium concentrations in the unknown samples are then determined from the established calibration curves.

The method is free from interferences with the exception of silver. Samples containing more than about 3 mg of silver must be treated for removal of this element prior to the initial fusion process or the final bead will no longer be a platinum bead but a variable platinum-silver alloy of unknown arcing properties. The removal is done as in the previous case, by reduction to silver metal and leaching with dilute nitric acid.

Pressure Dissolution — X-Ray Fluorescence Determination of Platinum Metals

R.E. Price

Inco Europe Limited, London, England

ABSTRACT

The pressure dissolution technique was developed by Gordon, Schlecht & Wichers and reported in 1944. It has been used at Acton Refinery since 1960 and is still the principal solution technique. Its principal advantages are efficiency, wide applicability and production of a clean solution free from alkali metals.

Up to 4 grams of sample may be treated with a wide variety of oxidising acid mixtures at temperatures of up to 270 degrees C and pressures approaching 4000 lbs/inch^2 in protected sealed glass tubes. Careful attention must be paid to practical and safety aspects, particularly the glass blowing techniques.

Recently, equipment designed and built by the U.K.A.E.A. has been acquired at Acton and has proved successful and safe.

KEYWORDS

Analysis; platinum metals; autoclave; pressure; dissolution; silver; gold.

INTRODUCTION

As you are all well aware, the platinum group metals have two outstanding characteristics as far as the analytical chemist is concerned. Firstly, they are very reluctant to be separated from one another and, secondly, they are difficult to get into solution. The former has been demonstrated in the preceding papers and I propose to discuss some aspects of the pressure solution approach to the latter characteristic.

Acid and oxidising acid mixtures are of limited value in opening up samples containing platinum group metals and more extreme techniques are often necessary, such as sodium peroxide or sodium bisulphate fusions, and smelting procedures followed by parting and cupellation operations. Useful though these methods are, they have a number of disadvantages in that they introduce unwanted elements such as alkali metals or materials from crucibles, volatile elements may be lost and other elements may be lost to slags or cupels. Often a second attack is necessary and this leads to a multiplicity of solutions. Alkali metals are a complication in both atomic absorption and x-ray fluorescence analysis.

137

The stability of the chloro- compounds of the platinum group metals suggested to Wichers, Schlecht & Gordon that the rate of attack on these metals by oxidising acid mixtures could be increased at temperatures above those normally attained at atmospheric pressure and led to their classic reports, published in 1944, in which they described a method of attacking difficult materials at elevated temperatures and pressures in sealed glass tubes.

Pressure solution has been used since then on a number of analytical problems involving difficult samples. Examples are refractory oxides, ceramics, tungsten and molybdenum alloys and radioactive samples, such as plutonium oxides, carbides and nitrides and, of course, the platinum group metals. The use of platinum lined sealed vessels for the decomposition of silicates in hydrofluoric acid has also been described. However, the technique does not seem to be as widely used as its potential analytical benefits might suggest and this is doubtless due, to some extent, to the practical and safety problems encountered in its use. I propose to deal with these aspects as we have encountered them at Acton Refinery rather than the theoretical aspects which are dealt with in the papers mentioned above.

In passing, I would refer to the small scale Juniper pressure vessels which are used to attack some materials at elevated temperatures. These are 50 ml PTFE lined stainless steel vessels with a screw clamped lid. They operate at temperatures up to about 140 degrees C. and pressures up to 250 psig. Our attempts to use these gave less than 10% dissolution of ruthenium and less than 5% of rhodium in aqua regia after two hours. Clearly, the operating temperature is not high enough for the platinum group metals.

At Acton Refinery equipment was constructed following closely on the designs described by Wichers, Schlecht & Gordon. The objective was to obtain complete solution of upwards of one gram of a wide variety of materials within a reasonable period of time. In practice, this is achieved overnight at a temperature of 265 degrees C. The oxidising acid mixture used is almost exclusively hydrochloric acid and chlorine and the reaction takes place in a sealed pyrex glass tube about 200 mm long and having an outside diameter of 19 mm and an inside diameter of 14 mm.

The construction of these glass tubes and the techniques of sample introduction and tube sealing are probably the most important factors contributing to the success and safety of the pressure solution method. Much depends upon the skill and experience of the operator and much attention must be devoted to his training. A stock of tube blanks is always available, having been made "at leisure". These are made by starting with a length of tubing twice the finished size and parting into two blanks. This is done by heating the central portion in a hot glass-blower's torch and drawing to a thin neck which can be broken to separate the two blanks. The seal is completed by heating the whole curved portion whilst rotating the tube using the thin section as a handle. Care must be taken not to pull or push this portion, so as to maintain a uniform wall thickness. Gentle blowing into the open end helps to produce a uniform round end, which is then annealed in the usual way. A neck about 2/3 - 3/4 of the tube inside diameter is then made at the other end, so as to provide a tube length of about 200 mm.

The sample, having previously been prepared as appropriate, is carefully introduced using a clean dry funnel, washing adhering material through with concentrated hydrochloric acid from a squeeze bottle. 10 ml of concentrated hydrochloric acid are added and the tube is transferred to a freezing vessel. This consists of an outer container about six inches tall and six inches in diameter, well lagged externally. It has a one inch hole centrally in the base with metal mesh liner supporting a central mesh cylinder about one inch diameter and six inches tall in which the tube is supported. The space between the container and the central cylinder is filled with coarsely crushed dry ice and the vessel is supported on a stand to allow free circulation of cold gas around the tube and out of the base of the vessel.

When the contents of the tube are completely frozen, the whole is transferred to a fume cupboard where 3 ml of liquid chlorine are added from a 10 ml measuring cylinder. The liquid chlorine is obtained from a standard $3\frac{1}{2}$ lb. chlorine cylinder inverted so that liquid is obtained from the valve.

Fig. 1   Loaded sample tube supported in the freezing vessel, together with an inverted chlorine cylinder showing the valve arrangement.

When the contents are frozen the tube, still in the freezing vessel, is transferred to a clamp and turntable unit arranged so that the portion of the tube above the neck can be clamped as nearly vertical as possible and the whole can be slowly rotated by hand. The clamp is fastened loosely at first to allow for expansion of the tube, and the neck area is heated for about thirty seconds with a luminous flame. After this period, the clamp is tightened more firmly and the flame is turned to full power, rotating the whole assembly by hand all the time. The neck wall collapses inwards and just before it seals the flame is removed, the clamp fixing released, and the top portion of the tube is raised rapidly. This gives a long neck which is cut off with the flame and run back on itself. A hot flame is maintained for a few seconds to make a smooth, even seal but longer heating risks the seal blowing out. A bushy flame, large but not quite luminous, is used to anneal the seal for about one minute, playing the flame over about one inch of the tube.

Fig. 2.   Tube being sealed while the turntable unit is rotated by the operator's left hand.

The freezing vessel and tube are then transferred to a safety cabinet and the tube is placed in a rack to thaw out. When the contents are liquid, it is inverted over moist litmus paper to test the seal. If it is not perfect, it can sometimes be re-sealed after re-freezing, but usually the trial would be re-started.

If the tube is satisfactory it is ready to be placed in a steel shell for heating. These shells are made in a similar fashion to those described by Wichers, Schlecht & Gordon. They are bored from solid hexagon bar, having an internal cavity 360 mm long and 28 mm diameter and lined with a glass fibre sheet. The open end of the shell is ground to a knife edge.

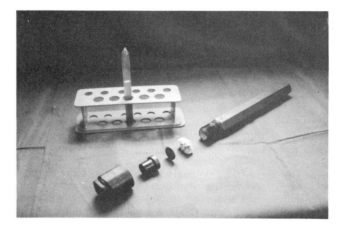

Fig. 3.   An exploded view of the shell assembly.

About 7 grams of calcium carbonate are placed in the bottom, followed by a glass fibre plug. The sealed sample tube is carefully introduced and the space around it packed with crushed dry ice powder so that the tube is held centrally and dry ice fills the space to the top of the tube. A glass fibre plug is added to fill the space to the top of the steel shell. Calcium carbonate is to neutralise acid in the event of a breakage and the carbon dioxide is to provide a compensating excess external pressure around the sample tube.

A copper disc is placed over the knife edge so as to provide a seal between the shell and the cap which is screwed to the shell by an external thread. This thread is treated with a molybdenum disulphide grease. The cap is screwed down and the effectiveness of the seal checked by immersion in water. It is necessary to keep the shells under water for about half an hour as leaks are sometimes slow in developing.

Six of these shells can be accommodated in the furnace assembly as shown in Figure 4 on the following page.

Fig. 4.   Six shells accommodated in furnace assembly.

They are arranged on a motor driven rotating wheel, so that they are at right angles to the wheel around its circumference and are carried around approximately horizontally at about twelve revolutions per minute. When mounted the six shells diverge to an extent of about half an inch from the axis of rotation so that, as they rotate, the contents of the tube both move around the tube and from end to end. An electrically heated tube furnace, closed at one end and mounted on wheels, is used to heat the shells. When in position the furnace is closed by the rotating wheel and it can be withdrawn to allow cooling and removal of the steel shells. Usual practice is to load the furnace in the afternoon and to run it overnight. It is switched off and withdrawn at about 6:00 am by overnight staff so that the shells are ready for opening when the day staff return. Thus the sample tubes have been at temperature (265 degrees C.) for about fourteen hours.

Fig. 5.   Opening the shells is carried out inside a safety cabinet
           on a steel bench provided with a special holder to accept
           the body of the steel shell.

The cap is "cracked" carefully to release the carbon dioxide and the glass tube is quickly transferred to the freezing vessel described earlier. When the contents are frozen the tube is scored about 1 - 1½ inches from the top and a red hot glass rod is applied to complete the crack. The top is carefully snapped off to be retained for washing and the tube and contents are quickly inverted into a one litre lipless beaker containing 25 ml 1:1 hydrochloric acid. A cover glass is placed on top and the contents of the tube are allowed to thaw slowly. Sometimes a quantity of liquid chlorine can be seen below the frozen plug of hydrochloric acid and as the temperature rises this chlorine can expel the tube contents rather violently. A thin jet of warm water from a squeeze bottle can be trickled down one side of the tube to create a narrow thawed passage to allow chlorine to escape. When conditions have stabilised, the tube and cap are carefully washed giving a sample solution of about 75 - 100 ml ready for analysis.

Although the usual sample weight used is one gram, up to three grams of material have been successfully dissolved, including iridium. For the majority of Refinery samples, which contain a wide variety of elements in varying quantities, a certain amount of sample preparation is necessary. Silica, lead and silver compounds and other materials will cause complications, either in the tubes or in the subsequent operations in a chloride media.

It is general practice at Acton to ignite all samples in silica crucibles followed by reduction under hydrogen. After cooling, they are transferred to a platinum dish and treated with hydrofluoric acid to remove silica. 5 ml of 1:4 nitric acid are added and the mixture is allowed to reflux gently under a watch glass for 30 minutes. After cooling, it is filtered through a 42 filter paper and the filtrate can be analysed as appropriate. After washing, the filter paper and insoluble are transferred to a silica crucible, ignited and reduced under hydrogen. The sample is then ready for the pressure solution process.

It is obvious that there are potential hazards associated with techniques using heated sealed tubes and attention to safety aspects cannot be over-emphasised. Operator training and experience are essential and appropriate safety clothing and equipment must be used. The tubes are comparatively safe to handle when frozen and should be taken to this state as quickly as possible. Probably the most hazardous stage is when the glass tube is being transferred from the steel shell after heating and it is essential that a safety cabinet or protective screen is used in addition to personal equipment.

Safety conditions led the United Kingdom Atomic Energy Authority Analytical Services Division at Harwell to develop a new portable pressure compensating autoclave.

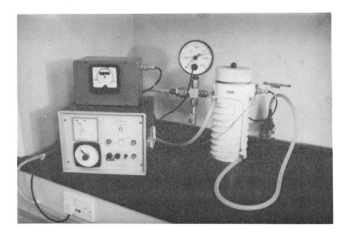

Fig. 6. Portable pressure compensating autoclave.

Samples containing plutonium compounds and other radioactive substances often require pressure solution techniques and clearly every care must be taken to prevent the escape of these materials.   This equipment consists of a stainless steel pressure vessel designed to operate at pressures of up to 6,000 lbs/in^2 and at temperatures of up to 320 degrees C.   It is electrically heated and equipped with water cooling and will accommodate up to four sealed tubes.   It uses silica tubes about 5 mm bore, a 2.5 mm wall thickness and about 200 mm long. A pressure gauge and a pressure release valve are provided, with a bursting disc set to limit the working pressure.   The temperature is controlled by a thyristor regulator and a clock is incorporated to switch off the heater at a pre-set time.   Devices are provided to switch off the heater if cooling water fails or if the pressure rises above a pre-set value and a visual alarm warns the operator.

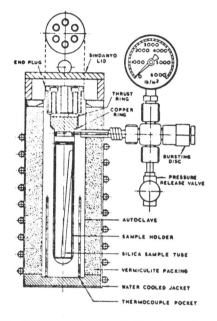

Fig. 7. Cross-section of the Autoclave.

As in the equipment described earlier, solid carbon dioxide is used to provide an external compensating pressure and calcium or sodium carbonate are added to deal with acid should a tube fail.   Up to four tubes are held in a sample carrier and the unit is sealed using a copper gasket with a sealing plug and thrust ring, final clamping being by six screws which are tightened using a torque wrench adjusted to 25 lb.ft.

Acton purchased this equipment in 1975 and it has been used without mishap.   Aqua regia was used as the oxidising acid mixture and up to 250 mg samples can be accommodated.   The tubes can be handled safely at room temperature after heating.   However, the small sample size is rather a limitation and this equipment requires operator attention during the initial part of the heating cycle to release excess pressure generated before the temperature stabilises.   Also, the handling of the small silica tubing requires different techniques from those used with the larger pyrex tubes.

Fig. 8. Materials dissolved at $320^{\circ}C$ and 4,000 lbf/in^2

Material and condition		Acid solvent	Heating period
Alumina	Ground sintered material	HCl	5 Hrs.
	Lump sintered material	HCl	10 Hrs.
	Calcined A.R.	HCl	15 Hrs.
Thoria	Fired at 2,000°C	HCl/HNO₃	5 Hrs.
Zirconia	Fused, ground to <100 microns	HCl/HNO₃	10 Hrs.
Magnesia	Powder A.R.	HCl/HNO₃	5 Hrs.
Ferric oxide	Fused	HCl/HNO₃	5 Hrs.
Ruthenium dioxide	Fine powder	HCl/HNO₃	5 Hrs.
*Plutonium dioxide	Heated in O₂ at 1,500°C	HCl	10 Hrs.
	Heated in argon at 1,550°C	HCl	16 Hrs.
	Arc melted	HCl	16 Hrs.
Iridium	Specpure-fine powder	HCl/HNO₃	5 Hrs.
Osmium	Specpure-fine powder	HCl/HNO₃	5 Hrs.
Rhenium	Fine powder	HCl/HNO₃	5 Hrs.
Ruthenium	Fine powder	HCl/HNO₃	10 Hrs.

*The Autoclave is used in an alpha-active handling facility for the dissolution of plutonium-containing material.

At Acton, pressure solution is used almost exclusively as the method of preparation of samples for x-ray fluorescence analysis. Final presentation to the spectrometer is in the form of a cellulose disc, starting with an aliquot of solution representing 0.1 g. from a 1 gram total sample taken to dryness and mixed with 1 gram of sodium chloride and 7.5 grams of cellulose. After careful homogenisation, the mixture is compressed under a hydraulic pressure of 30 tons to a disc of controlled size suitable for the spectrometer sample holder. A Philips 1220C computer controlled spectrometer is used and this cellulose disc technique allows synthetic standard discs to be made up covering a wide range of elements and compositions. These are used to calibrate the spectrometer, providing inter-element effect data to be calculated and used by the computer. Thus, all samples are taken to the same physical form and, as closely as possible, the same chemical form as the standards - the objective of the sodium chloride addition being to allow for the fact that the chloride content is difficult to control through the preparation procedures. In practice, having obtained an analysis of a sample from the spectrometer, a matching synthetic disc prepared from standard solutions and compared with the sample disc.

Pressure techniques in sealed tubes at elevated temperatures produce solutions suitable for subsequent analysis by most methods. These solutions are generally free from contaminating elements from crucibles and other vessels and the chloride media and absence of alkali metals makes them particularly amenable to atomic absorption and plasma techniques. A wide variety of intractable materials may be treated in this way and, in the hands of well trained and careful operators, can be used safely in both research and production control laboratories.

REFERENCES

Wichers, Schlecht & Gordon, Research Papers 1614 and 1621, Journal of Research of the National Bureau of Standards, Volume 33, 1944. Attack pf Refractory Platiniferous Materials by Acid Mixtures at Elevated Temperatures, 363-381, 451-470.

Metz & Waterbury. Los Alamos Scientific Laboratory, U.S.A. Report LA-3554(1966)

Economics

Excitement in the Metals Markets — A Banker's View

P.C. Cavelti

The topic of my speech today is "Excitement in the Precious Metals Markets - A Banker's View". I'm very fortunate to have an audience which needs no introduction to the activity in precious metals markets over the past ten years or, more specifically, to the wild advance in the gold price from around $250 one year ago to $850 in early January of 1980. When gold prices climb to new record highs it is inevitably linked to bad news of some kind. The investor, who often does not understand this relationship, senses that something in our economic, political or social structure is wrong and turns to his banker for advice. The most recent rush in metals prices made gold and silver household words to more people around the globe than ever before. The headlines and coverage in the media were inevitably spectacular and most private investors were fascinated by its appeal but regarded it as something rather exotic and speculative. Many newspapers and magazine reports pointed more towards the mystic element in gold, which did not help people in making an investment decision. The question posed to their financial managers or bankers was inevitably: "What is gold?" And whether gold can or cannot be advocated as a component of a private investment portfolio has been more widely discussed during the past twelve months than ever before.

What is gold really? We all know that gold is a non-rusting, malleable, ductile, soft metal with a low melting point and a high density. These qualities make it an outstanding metal for which our economic and industrial system has some real uses, which is an important factor, because it makes for a natural flow of demand and supply and it creates the basis for anything that is to serve as a store of wealth.

However, as a financial manager, I am equally interested in the traditional aspects of an investment. And when I think about those what comes to my mind first is probably negotiability. Gold bullion certainly is highly negotiable in that any of us could go to one of the leading gold dealers in Toronto today and buy one ounce of gold and then sell it against cash tomorrow morning in either London, Zurich, Hong Kong, Frankfurt or New York.

However, negotiability is not the only investor requirement, although it is possibly the most important criterion. So let's examine what else makes gold a valid investment. A vital factor is distribution. If an investment is not well distributed, then it becomes subject to manipulation. In the case of gold, the geographic flow alone indicates an excellent distribution. In Table #1, which is reprinted with the permission of Consolidated Gold Fields Ltd., you can see how gold leaves its production centres around the world and finds its way to the marketplaces which can generate industrial and investor demand. It is held by governments everywhere, it is kept by multi-national corporations, it serves banking groups, large private wealth and a multitude of small investors in a myriad of nations. One could indeed say that gold is very well distributed.

The banker and investment manager also has to concern himself with the cost of turning an investment over. The comparison on Table #2 indicates that turnover costs in major gold market centres are almost uniform. I have deliberately used a small unit, that of a one ounce bullion wafer or a Krugerrand. As you will note, it costs somewhere between three and four percent net to purchase and sell a Krugerrand. Whether this is done in Frankfurt, Hong Kong, Toronto or Zurich is immaterial. (Sales taxes, which apply in some jurisdictions have not been included in this comparison.) If we switch over to the difference between buying and selling prices for a regular one ounce bullion wafer, the turnover cost becomes even less. In the comparison of major international market centres, an average turnover cost of exactly 2.5 percent is reflected. It appears that Canadian investors are even more fortunate than their counterparts elsewhere. If you purchased a one ounce gold wafer in Toronto today, you could sell it back tomorrow with a loss of less than 1.7 percent. That is, of course, if the price of the metal did not change in the meantime. Seen from this angle, gold positively qualifies as an investment. There are very few vehicles anywhere, which allow you to turnover a $500 or $600 investment at such little cost.

These three main points, distribution, negotiability and low transaction charges bring gold into line with other investments today's financial manager can recommend. This, however, is where the similarities end.

To begin with, gold does not yield any income, a drawback most investors find very hard to accept. Secondly, there is a storage risk involved and, thirdly, many people feel almost guilty to invest in gold. Bullion has been dubbed a "Profiteers' Vehicle" or an "Investment in Doom", particularly in North America. The U.S. ban on private gold ownership and the aggressive demonetization rhetoric employed in the late 70's have left their scars.

But these three negative points are easily outweighed by gold's most popular feature: that it moves counter-cyclically to economic growth **and wealth and** that our political, social or economic structures are in trouble gold edges up in prices. To me, this "insurance aspect" is really what makes gold an excellent alternative investment. There is ample evidence in all free economies in this world to prove that gold can be relied upon to stay ahead of price inflation and to protect wealth against economic or other perils. Let's look at the graphs on Table #3. The first graph shows how gold has stayed ahead in terms of price inflation. I have charted U.S. wholesale commodity prices from 1780 to today. During this two hundred year period, the price of gold in U.S. dollars has stayed solidly ahead of the price of a basket of goods denominated in the same currency. The only deviation from that pattern was during a period when gold prices were artificially controlled.

TABLE 1

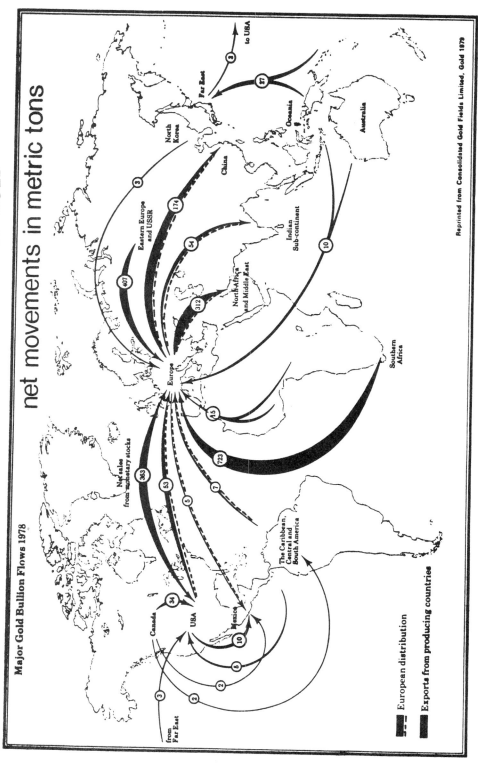

THE FLOW OF GOLD

net movements in metric tons

Major Gold Bullion Flows 1978

Reprinted from Consolidated Gold Fields Limited, Gold 1979

European distribution

Exports from producing countries

TABLE 2   TYPICAL RETAIL SPREADS FOR SMALL GOLD BULLION UNITS – WORLD WIDE

CITY	DEALER	DATE	DOLLAR PRICE PER KRUGERRAND	% DIFFERENCE BUY / SELL	DOLLAR PRICE PER 1-OUNCE UNIT	% DIFFERENCE BUY / SELL
Frankfurt	Dresdner Bank	May 15/80	521/539	3.5%	515/528	2.5%
Hong Kong	Hang Seng Bank	"	516/536	3.9%	512/528	3.1%
London	Mocatta & Goldsmid	"	519/536	3.2%	513/524	2.2%
New York	Republic National Bank	"	524/542	3.4%	514/526	2.3%
Toronto	Guardian Trust	"	522/541	3.6%	514/523	1.6%
Zurich	Credit Suisse	"	521/539	3.5%	513/529	3.2%

TYPICAL RETAIL SPREADS FOR SMALL GOLD BULLION UNITS – CANADA

CITY	DEALER	DATE	DOLLAR PRICE PER KRUGERRAND	% DIFFERENCE BUY / SELL	DOLLAR PRICE PER 1-OUNCE UNIT	% DIFFERENCE BUY / SELL
Toronto	Bank of Commerce	May 15/80	519/538	3.7%	514/522	1.6%
Toronto	Guardian Trust	"	522/541	3.6%	514/523	1.8%
Toronto	Scotiabank	"	524/539	2.9%	514/522	1.6%

TABLE 3

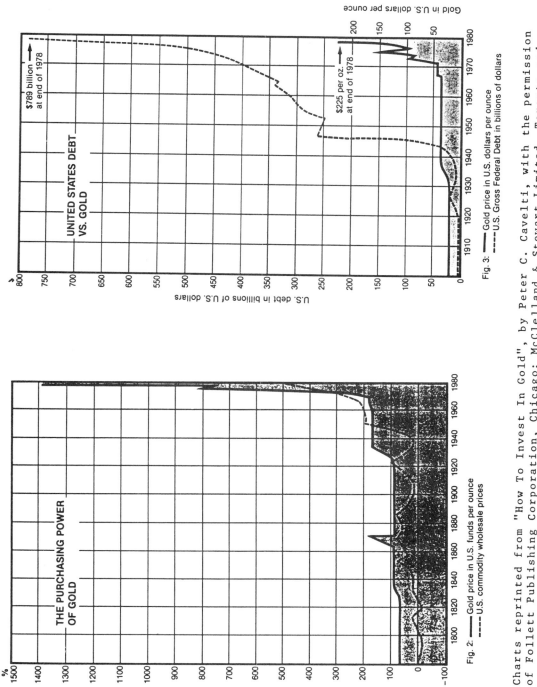

Fig. 3: —— Gold price in U.S. dollars per ounce
         ---- U.S. Gross Federal Debt in billions of dollars

Fig. 2: —— Gold price in U.S. funds per ounce
         ---- U.S. commodity wholesale prices

Charts reprinted from "How To Invest In Gold", by Peter C. Cavelti, with the permission of Follett Publishing Corporation, Chicago; McClelland & Stewart Limited, Toronto and Maximus Press Limited, Toronto.

In the second graph you can see that there is a relationship between the creation
of debt and the price of gold. Again, we use the United States economy as an
example. The chart illustrates how the United States gross federal debt increased
during the years of the second world war and the subsequent period of reconstruc-
tion. The price of gold bullion was frozen during that period and was only
allowed to float freely in the early 1970's. How quickly it corrected back you can
for yourself. Between 1970 and 1979, the U.S. debt doubled from $400 billion to
$800 billion. During the same time, the gold price more than quadrupled, moving
from $44 to $225. During 1979, a period not reflected in our graph, the federal
gross debt rose to approximately $840 billion, growing at an annual rate of 6.5
percent. The gold price, as you all know, moved to around $525, an astounding
increase of 133 percent!

The reason I have used the U.S. economy as an example of gold's ability to stay
ahead of price developments and the creation of debt is simply because the United
States economy is the one mechanism that affects us all. However, a study of the
gold price in relation to a group of major currencies and their economies is even
more convincing. Recent history provides many examples, but let us use the period
between 1977 and 1979 for our study. It is a time we all remember for its
volatility in currency and precious metals markets. You may remember that the
United States dollar came under considerable pressure during 1977, a trend which
reached alarming proportions in 1978. During this period the central banks of
Europe and Japan were once again faced by an old dilemma. Should they let the
value of their currencies rise until their export competitivity was threatened
or should they support the United States dollar at the risk of "importing"
inflation? Following a rather inconsistent course, the majority of these nations
finally showed significant increases in their money supply in the third and
fourth quarters of 1979. Consumer prices were also correspondingly higher than
had been the case before. On Table #4, you see a comparison of the key indus-
trialized western nations and indicators characterizing their economies. The two
I would like you to concentrate on are marked with an arrow, namely the money
supply changes and the consumer prices. On the money supply side you see how
M1 shows a twelve month increase of 6 percent in the case of West Germany, 10.5
percent in the case of Japan and 13.76 percent in the case of Switzerland!

Surprisingly, all of these increases were larger than that recorded by the United
States of 4.7 percent. Similarly, the inflation rates started to heat up. Where
inflation rates had been below 2 percent just a few months before, they were now
5.3 percent in the case of West Germany, 3.1 percent in Japan and almost 5 percent
in Switzerland! This followed a pattern, which had been established within the
U.S. where the inflation rate had shot up to almost 12 percent during the previous
few months. What all this did to the price of gold is illustrated on Table #5,
where you see the bullion price charted against the world's major six currencies.
You see how during 1977 and 1978 the gold price moved upward sharply against the
Canadian dollar and the United States dollar. It did not do too much in terms of
the German mark and the Pound Sterling and it literally stayed even against the
Japanese and Swiss currencies. But during 1979, as inflation was exported to
virtually every economy on earth, prices in all currencies started to gain
drastically. At the end of that year the gold price in Canadian and United
States dollars reflected a three year increase of around 300 percent. In terms
of the Pound and the yen it was around 200 percent. Against the German mark, gold
had gained 170 percent and even in terms of the mightly Swiss franc gold had
advanced 140 percent!

TABLE 4　econochart

DATA	BRITAIN	CANADA	FRANCE	W. GERMANY	ITALY	JAPAN	NETHERL.	SWITZERL.	U.S.A.
POPULATION in millions	56	23	53	61	56	114	14	6	217
REAL GROSS NATIONAL PROD. nominal, in US$ billions, 1977	245	192	381	516	195	691	107	63	1,887
GNP PER CAPITA in US$, 1977	4,378	8,297	7,150	8,408	3,461	6,066	7,704	9,971	8,704
REAL GNP GROWTH percentage changes, 1977	+ 1.6	+ 2.7	+ 3.0	+ 2.6	+ 1.7	+ 5.2	+ 2.5	+ 2.7	+ 4.9
1978 1979	+ 3.0	+ 3.5	+ 3.0	+ 3.0	+ 2.0	+ 5.75	+ 2.5	+ 1.5	+ 3.9
TRADE BALANCE in national currency (in mio.) 1978	-1,104	+3,520	+2,029	+40,700	-348	+US$20,845	-6,469	-804	-28,860
previous month	- 100	+ 18	-3,170	+ 607	-388,000	- US$669	-639	-680	- 1,110
latest month	SEP - 100	AUG -141	SEP-1,790	AUG+ 976	AUG+583,000	SEP+US$140	AUG-1,200	SEP - 445	AUG - 2,360
EXPORTS, % OF GNP 1977	23.8	19.5	17.1	22.8	23.0	11.7	40.9	27.7	6.4
IMPORTS, % OF GNP 1977	26.4	16.6	18.5	19.6	24.3	10.3	43.7	28.3	8.3
OFFICIAL RESERVES in US$ millions, latest month	SEP22,750	SEP4,510	SEP38,946	SEP56,784	AUG 36,200	SEP 25,330	SEP13,049	SEP23,624	AUG 20,020
MONEY SUPPLY CHANGES (M1) latest month	AUG - 0.1	SEP +0.3	JUL +0.1	AUG +0.4	MAY +1.4	AUG +0.8	JUL -4.7	JUL -0.6	SEP +0.9
12 mos. to date	+11.5	+7.4	+12.3	+6.0	+22.8	+10.5	-0.1	+13.76	+4.7
UNEMPLOYMENT (ADJUSTED) In 1,000's at year end 1977	1,422	1,154	1,070	1,023	1,501	1,180	205	12	6,310
In 1,000's at year end 1978	1,320	1,230	1,288	949	1,594	1,170	211	13	6,000
Latest month in percentage	SEP 5.2	AUG 7.2	AUG 5.9*	SEP 3.6	AUG 7.4*	JUL 2.3	AUG 5.0	AUG 0.3	SEP 5.8
Year ago in percentage	5.8	8.5	5.2*	4.3	6.8*	2.3	5.2	0.3	6.0
CONSUMER PRICES percentage changes 1977	+ 15.9	+ 9.5	+ 9.5	+ 3.9	+ 17.0	+ 8.1	+ 6.4	+ 1.3	+ 6.5
1978	+ 8.0	+ 8.4	+ 9.5	+ 2.5	+ 13.5	+ 4.0	+ 4.5	+ 1.1	+ 7.5
latest month	AUG+ 0.2	AUG+ 0.4	AUG+ 1.0	SEP + 0.1	SEP + 2.5	SEP + 2.5	SEP + 0.9	SEP + 0.5	AUG + 1.1
12 mos. to date	+ 15.8	+ 8.4	+ 10.8	+ 5.3	+ 16.8	+ 3.1	+ 3.9	+ 4.8	+ 11.8
EXCHANGE RATE against US$, today	2.1450	0.8494	0.2362	0.5545	0.00120	0.00427	0.5040	0.6040	-----

TABLE PREPARED October 17, 1979　　RESEARCH PROVIDED BY GUARDIAN TRUST COMPANY:　INTERNATIONAL DEPARTMENT

TABLE 5   GOLD'S   GLOBAL   RISE: 1977 — 1979

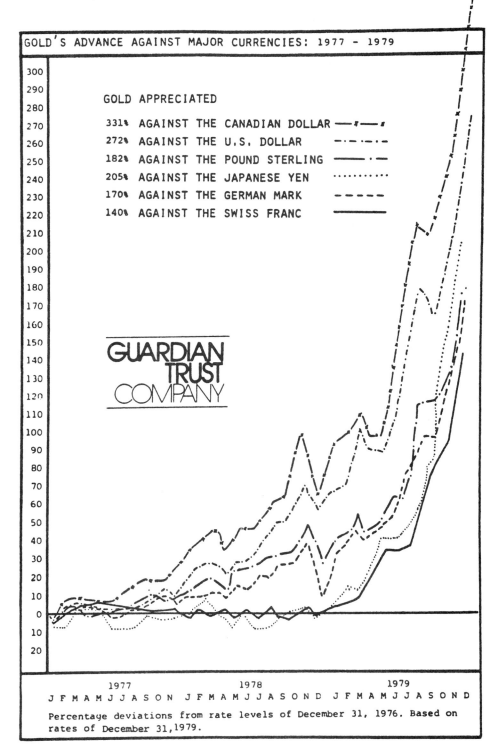

GOLD'S ADVANCE AGAINST MAJOR CURRENCIES: 1977 - 1979

GOLD APPRECIATED

331% AGAINST THE CANADIAN DOLLAR —·—·
272% AGAINST THE U.S. DOLLAR —·—·—·—
182% AGAINST THE POUND STERLING ——·—
205% AGAINST THE JAPANESE YEN ···········
170% AGAINST THE GERMAN MARK ———————
140% AGAINST THE SWISS FRANC ————————

Percentage deviations from rate levels of December 31, 1976. Based on rates of December 31, 1979.

This ability to provide protection against the perils of inflation is unique. There simply are no traditional investment alternatives, which have established track record in maintaining their value over an extended period of time in an inflationary environment!

But having pointed out all these qualities and advantages, which make gold acceptable as an investment and highly endorsable as an insurance or protective asset, we should have a look at its price today. Is gold a valid insurance at current price levels? Should those who did not buy at $35 or $200 go out today and buy gold at $550? To the person interested in short term gains I would say no. The past year has shown us how treacherous and volatile gold markets can be and I have similar expectations for the period to come. However, to the longer term holder who is interested in security I continue to recommend gold holdings at current price levels.

Today's world is characterized by rapid and powerful changes. More monetary wealth is concentrated in one particular geographic area than has been for a long time. In addition, it is that very area which produces the commodity we so desperately depend on, oil. No one needs to stress that any further deterioration of the political balance in the Middle East could cause capital shifts of unprecedented size. They would affect the values of gold, currencies or any other financial markets very quickly and in a very major way. Then, there is the economy. The problem of inflation is unlikely to disappear, because we remain highly dependent on expensive foreign energy sources and because a majority of key governments cannot or do not want to face the social consequences of abandoning their habits of fiscal indiscipline. Our international monetary system also faces problems and then there are serious questions concerning the structure of international debit. All of this, ladies and gentlemen, makes for an environment of high uncertainty. During the time of such uncertainties it is not possible for anyone to pick an investment vehicle which is sure to perform well. The variables and possible outcomes of today's situation are simply too many. However, one thing you can do during such uncertain times is to pursue the right investment strategy. A strategy that will not necessarily produce the highest gains, but will grant a very high degree of protection is diversification. And, in my opinion, a diversified investment approach cannot afford to ignore gold bullion. As an insurance asset, I recommend that between ten and twenty percent of an investor's overall worth should be held in gold. How much exactly this should be depends on the investor's personal objectives, his potential liabilities, his family situation, his age, his attitude towards risk and the degree to which the current political and economic changes worry him.

Today's gold markets offer a vast array of different investment vehicles. Bullion, certificates, futures contracts, mining shares and numismatic coins are just a few. All have their individual merits and drawbacks. To the security-minded, long term insurance-conscious investor I recommend gold only in the form of fully-paid bullion or bullion type coins, such as the Canadian Maple Leaf or the South African Krugerrand. Only gold in this form, manufactured by an internationally recognized refiner or by a major government, offers low transaction charges, negotiability and recognition in all important market centres, and is portable and liquid at all times.

Platinum Group Metals — Commodity Studies

T. P. Mohide

Director, Mineral Resources,
Ministry of Natural Resources, Ontario Government,
Whitney Block, Queen's Park,
Toronto, Ontario, Canada

ABSTRACT

This paper outlines Ontario's new comprehensive world wide review and commodity study on the platinum metals, covering sources and uses and related government policy.

KEYWORDS

Platinum, Palladium, Rhodium, Iridium, Ruthenium, Osmium, Uses, Commodity study preparation, government policy, value of mine production of precious metals, world platinum consumption, Ontario - largest mine output of precious metals in North America, Canada, Soviet output, South Africa.

In this session, I follow on behind such internationally eminent authorities as Simon Strauss and Irwin Shishko that I really hesitate to say anything at all. I have personally learned much from them over a period of many years, as have many others in the business. I am therefore extremely honoured by your invitation to speak today.

First let me say something about Ontario and its metals.

At its furthest points, Ontario is about one thousand miles east to west and one thousand miles north to south. In area, it is about the size of Texas and California put together! Virtually all of Ontario's border with the U.S.A. is water and is about 1,715 miles long, fronting on five States of the U.S.A., Minnesota, Michigan, Ohio, Pennsylvania and New York State.

The value of Ontario's metallic mineral output is equivalent to almost half the value of the entire United States output from their mines. This surprises many people. That's just Ontario, not the whole of Canada. So you can see that we mine, smelt and refine metals in a very big way.

Ontario is the world's largest single source of nickel and is also a
major mine source of copper, zinc, lead, cobalt, selenium, precious
and other metals.

In terms of value, Ontario is the largest single mine source of precious metals in North America. The value of Ontario's mine output of
precious metals in 1979 is estimated as about 500 million dollars.
Over a thousand dollars an ounce for platinum earlier on this year,
as well as strong gold prices, will help to boost the 1980 value of
our mine output of precious metals.

In the eight precious metals, Ontario enjoys a very special world prominence.
1) Canada has been the world's largest mine producer of silver in
    most of the years since 1967 and the Province of Ontario is easily
    the largest producer of silver here.
2) Canada continues to be a leading world mine producer of gold and
    Ontario remains the largest producer among the Provinces.
3) Ontario is the world's third largest mine source of the six plati-
    num group metals after South Africa and the Soviet Union. These
    metals come mainly from the Sudbury district about 200 miles north
    of here. These three major sources make up about 98 per cent of
    the total world platinum group metals output. Platinum, we are
    happy to note, is almost always much more expensive than gold.

Our Ministry studies show that Ontario mines will continue to produce
such large quantities of the eight precious metals well into the 21st
century.

These metals are therefore among those in which the Ontario government
has a great interest. My responsibilities include that of ensuring
further growth in mining, smelting and refining the resulting metals
in this Province. We certainly extend a hearty welcome to further investment by free enterprise in our mines and metal operations and in
attractive potential mineral development situations, which include
precious metal opportunities.

The lack of a comprehensive overview of the sources and of a clear
picture of the place of these platinum metals in the world and the onset of dramatically rising prices were some of the reasons why the Ontario government decided that the time had come to add the platinum
group metals to its growing list of mineral policy background papers
on mineral commodities, in which we publish a substantial part of our
findings on mineral economics and metal matters.

Owing to pressure of other business, this study, "Platinum Group Metals - Ontario and the World", a comprehensive world wide review, was
written in my spare time, (although no one believes this), in eight
months and was printed in the second half of 1979. I had been a co-
author of books before, but this time I was on my own.

The study is not intended solely for the highly technical reader, but
is aimed at a much wider audience.

Platinum is an extremely elusive subject. As you know, the platinum
group metals are poorly documented in many respects, taking the broad
view. This reflects perhaps the general penchant for secrecy in the
world of precious metals which is perfectly understandable from many
points of view.

When doing a commodity study on metals such as zinc or copper, the first problem is that there is too much data available to digest. With platinum metals, the trouble is that there is too little data, which therefore has to be assembled piece by piece.

You have to proceed to ransack every piece of information on platinum you are able to lay your hands on and question every expert you can get hold of and then get it all into writing.

The process if pedestrian and not at all scientific - no doubt to the dismay of many here today. Rather it is something like assembling pieces one by one in order to produce a mosaic, knowing however that one will never obtain a crystal clear picture, but only an impressionist one. Platinum remains a world of incomplete facts and all conclusions are coloured accordingly.

However, it was certainly fun finding out. In spite of many years as a trader in precious metals and later as Chief Executive of a Commodity Exchange, I kept finding more and more areas of my own ignorance.

As a government official, one is no longer in the hurly burly of the market place in which platinum group metals move. One sees metals through a different telescope. Industry sources are often sealed off as company confidential or in one way or other the desired data is not obtainable. There is sometimes, for example, a natural fear in industry that a government might "leak" a piece of confidential information, even accidentally.

There is of course also a serious danger that a government writer could be so cut off from industrial reality that the resulting paper would be "ivory tower" or maddeningly theoretical. He does however enjoy the advantage of being able to take a detached view.

As one proceeds, the contents grow slowly as, facet by facet, the picture emerges. So does the list of items under the heading "no information" or "no data available". The gaps appear alarming.

The target date for publication was set, so one was denied the luxury of "all the time in the world". Nevertheless, as many of you know, one's best work is often produced under such pressures.

A time-consumer in doing commodity studies is "conflict of data". Estimates by distinguished observers over the years on subjects such as "world platinum consumption" were not and are not in harmony. Very wide differences exist.

The uncovering of uses of the various individual platinum group metals which were new to me was a fascinating experience.

One was struck again by the curious fact that the mines that produce virtually all of the platinum group metals in the world are located outside of the countries where there is the greatest consumption of them, that is Japan, U.S.A. and Western Europe. However, the Ontario mines are only about 100 air miles from the U.S. border.

Enjoyment increased as one got into the fascinating question of the future - what is the demand for these metals going to be in ten years time, twenty years time? Where will the metals come from? What mines? Will there be further stockpiles - and so on. Just about everyone wants to know about the future. What will happen to the South African platinum mines in a rapidly changing Africa? What will happen at the Soviet mine complex at Norilsk that produces almost all of their platinum group metals?

The commodity study goes into world mine production, refining, stock-piling and the various markets for these metals, as well as the uses of platinum in jewellery of all kinds, but it does much more than that. It goes on to describe other less well known but equally fascinating and vital uses of platinum - many in the world of medicine - for example in heart pacemakers and anti-cancer drugs.

There are many uses in surgery and as a special no-chemical pain kil-as well as in dentistry, in alloys and in making very powerful magnets for artificial eyes and teeth. It is also used in producing such diverse items as orchids and jet engines, in making television glass, in producing fibre-glass, in manufacturing rayon, in the refining of gasoline and in your automobile's exhaust purifier that you probably never see.

One extremely important use of platinum is in the production of arti-ficial fertilizers which greatly increase world food output. This enables Canada and the U.S.A. to produce such large surpluses of grain, which help to feed not only us, but many people and animals around the world, preventing famine and starvation in certain countries.

Platinum is also vital in your toaster, your electric shaver, in mak-ing many of the fabrics that men and women wear, in keeping food fresh in store over long periods, in the production of laser beams and so on.

It is fascinating to note that new uses for platinum are being found all the time but the older uses do not die away. One major new use is in producing electricity on a large scale from fuel cells, without pollution or noise. The result of these old and new demands has been that total world demand for platinum group metals has rocketed to five times what it was about twenty years ago and prices are at record le-vels.

The paper also describes the uses of the other less known but very im-portant metals of this group which are also mined in Ontario and a few other countries. Ontario is the world's third largest producer of these other platinum group metals.

Briefly, some of these are:

Rhodium   -the world's most expensive precious metal
           -vital (with platinum) in making artificial fertilizers for
            greater food production
           -vital in fibreglass production
           -in costume jewellery
           -in expensive mirrors

Palladium -vital in your telephone systems
          -used in making antibiotics
          -used in making "white gold", called the poor woman's plati-
           num by some

Iridium   -the world's third most expensive metal
            -used in producing laser beams
            -vital in jet engines
            -used in making surgical scalpels extremely sharp
            -used with platinum in heart pacemakers

Osmium   -a cubic foot of osmium weighs about two thirds of a ton
            -used in electron microscope work as a special stain

Before a government can consider producing a policy, particularly mineral policy, the subject must be massaged properly. The salient facts
must be assembled from around the world and a thorough analysis prepared before any array of options for policy can be produced for review by the Minister, who is an elected member of the Ontario Legis-
lature and a member of the Cabinet. In the study, such policy options
are offered to the Government of Ontario.

Minting

The Gold Maple Leaf

Denis M. Cudahy
Vice-President Manufacturing
Royal Canadian Mint

In 1970, the Mint was split off from the Department of Finance and established as a profit-oriented Crown Corporation. It operates under the direction and control of a Board of Directors appointed by the Governor in Council. The Board is composed of three members from the Public Service of Canada, three from the private sector, and the Master of the Mint. Under this kind of organisation, the Mint has prospered -- in 1970 revenues totalled $8 million and by last year had increased to $800 million. Next year, I expect revenues will exceed $800 million.

The change from 1970 to 1978 was the result of an aggressive marketing and production effort to obtain contracts for the production of coins for 34 countries, and the opening of new markets for our numismatic coins outside of Canada -- we sold more than $15 million in Europe last year. The metals purchased from private industry were extracted mostly from Canadian mines. In 1978, we used 100,000 ounces of gold, 300,000 ounces of silver, 4.2 million pounds of nickel, 12 million pounds of copper, 1.6 million pounds of stainless steel, 2 million pounds of aluminum, plus other metals, for a total value exceeding $55 million.

And now we are seeing a more dramatic change. With the introduction of the new bullion coin program, an additional two million ounces of gold will be purchased in 1980, for a total value of over $700 million.

Enough statistics and general information about the Mint. The topics that I think will be of particular interest are a review of the policy decisions that led to the minting and marketing of the new Gold Maple Leaf, as well as the background leading up to the establishment of the technical specifications for the bullion coin, the packaging material, as well as a review of the production operation.

165

THE ORIGIN OF THE BULLION CONCEPT

The concept of bullion coins originated in South Africa in the nineteen-sixties when the Chamber of Mines was asked to become involved in the marketing of gold, so as to offset rising production costs and a fixed price for gold.

In 1964, following several years of meetings and discussions, the idea of producing a gold coin containing one ounce of gold but no face value was generally accepted. It was the best answer to the problem posed by a gold coin restricted to a face value while subjected to the fluctuating value of the metal content. The legal tender status was then sought and, following lengthy negotiations with representatives of the Government, the South African Mint Coinage Act was amended in October 1966 to include a new currency, the Krugerrand, containing one ounce of gold, as part of their monetary system, but not directly related in value to the official currency, the rand.

The first coin was produced by the South African Mint and sold to the public by the Chamber of Mines in July 1967, and six years later sales reached three million coins. By 1978, over 6 million coins had been sold.

THE GOLD MAPLE LEAF

The approach to Canada's first gold bullion coin was slightly different. As you know, we have been minting numismatic gold coins since the early seventies for other countries and, since 1976, for Canada. However, the bullion coin presented a totally new challenge for us, as well as for every private concern involved in the process.

In mid-February of 1979, the Government authorized the Mint to proceed with a bullion coin program on a trial basis for a period of three years, during which time five million coins would be minted and sold to the public. By mid-June 1979, we were in full production and by mid-August we began distributing coins in preparation for the launching of the program in September 1979. To date, the program has been a success and it is due in large part to a number of well known individuals, not all Canadians.

When the price of gold dropped dramatically from a high of $185 at the end of 1975 to a low of about $104 in 1976, there was cause for concern in the Canadian gold mining community. The Canadian gold producers met at that time with various levels of the Government to explore the possibility of putting their gold production into coinage to back up Government loans needed to support a depressed industry. While the reaction of the Government was not very positive, the idea did not die because of the success of the Krugerrand. It was revived again in 1977 when Herbert Coyne, a well known U.S. gold trader, is reported to have said to John Lutley, a Canadian industrialist: *"Why don't you Canadians get into the bullion coin business?"*

As a founding member of the newly born Gold Institute, John Lutley not only set out to find the answer, but he pursued the proposition vigorously. In fact shortly thereafter, John approached the Mint to see if the Government would be receptive to the idea of producing a gold bullion coin.

This resulted in several meetings to discuss not only the details, but the course of action needed to pursue the idea of a Canadian bullion coin. It was soon established that it would be necessary to obtain the agreement of the Minister of Finance and the Receiver General for Canada (better known as the Minister of Supply and Services), both responsible for the implementation of the Currency and Exchange Act, and that special legislation would be required. In all, not an easy task. However with the input and interest of many people which included the Mining Association of Canada and its Gold Committee, as well as proponents from the private sector, it was possible to convince all concerned that Canada should be fabricating its raw materials and making finished products whenever possible. They felt that a gold bullion coin program, in addition to stimulating the Canadian gold mining industry, would create new jobs, help curb inflation, and make a significant contribution to Canada's balance of payments.

In May of 1978, after much enthusiastic debate in Parliament, an amendment to the Currency and Exchange Act was approved, thus enabling Cabinet to authorize the minting of gold coins. The amendment received Royal Assent on June 30, 1978.

Following Royal Assent, an interdepartmental committee of senior officials representing the Department of Finance - Energy, Mines and Resources - Industry, Trade and Commerce and the Royal Canadian Mint began an assessment of all the economic considerations and financial implications such a program would entail. One of the tasks of the committee was to analyze the benefits perceived by the gold mining industry and evaluate the impact of the proposed program on the consumption of Canadian-mined gold; the opening of new markets for the sale of newly mined gold; the creation of a significant number of jobs; the increase in gold production and exploration; the price of gold on the world's market; and the Canadian economy.

After several months of study, the committee concluded that the specific benefits mentioned could not be corroborated, based on the documents provided by the private sector or information already available in various departments of the Federal Government. Nevertheless, it was recommended that a gold bullion coin program be approved on a trial basis in order to ascertain whether or not there were benefits to be derived by Canada as a whole and the mining industry in particular. It was further recommended that such a program be launched for a period of three years and the benefits assessed during that time.

Following the committee's report, which was supported by the Canadian gold mining industry, the Government approved the Canadian Gold Maple Leaf Program in February of 1979. All of this could not have been achieved without the initiative and hard work since 1976 of the Canadian gold producers.

While the above discussions were taking place, the Mint was evaluating various proposals as to the purity, size and design. In addition, the process considerations as well as cost implications were being reviewed in relation to the purity, size and design.

          GOLD PURITY

There were a number of discussions during 1978 as to whether we should consider only a one ounce coin or whether half and quarter ounce should also be considered.

However, it was decided, after reviewing the sizes with experts in bullion transactions that we would be able to sell the limit, that a half ounce may lead to confusion with a new program introduction and that a one ounce precious metal coin appeared to be the most popular throughout the world.

We also had to consider the confusion which may have been generated in relation to our half ounce numismatic coin even though each was destined for a different market.

Following this, we had to decide whether this one ounce gold was to be a nominal
one ounce coin or whether we would ensure that each coin contained at least one
ounce of fine gold (i.e. more than 1 ounce of fine gold in each coin). It was
finally decided that we would guarantee the one ounce of fine gold content.

Consideration was also given to whether we should issue a 900.0 gold, 916.7 or
pure gold coin. However, it was recognized that if an alloyed coin was to be
produced, we had to decide on an alloying element. As you know, there are a
number of elements which could be considered with the more traditional ones
being: copper and silver.

Since we had had a good reception to our numismatic half ounce gold coin when
alloyed with silver, we felt that only silver should be considered if we were to
obtain the delicate yellow of pure gold and still have a precious metal coin. We
had found that the delicate yellow colour was well received and very popular with
those who bought the half ounce numismatic coin.

Inquiries were also made in various parts of the world as to the importance of
the colour of the bullion coin and the responses supported the position that the
more closely the coin resembled pure gold, perhaps the more attractive it would
be to the final purchaser. Therefore, we limited our search for an alloy to
silver. However as we progressed, it was obvious that if silver was used, the
cost of silver would be prohibitive since no cost recovery could be obtained for
the silver content. It was, therefore, tentatively decided to issue a pure gold
coin.

Before finalizing our decision, we evaluated quotations for the supply of coin
blanks from private industry for both alloyed blanks and pure gold blanks. There
was no cost penalty for pure gold and in fact, a cost benefit due to the ease of
processing.

The blank suppliers did caution against pure gold as they felt that they would
have difficulty in obtaining the required hardness and that something else should
be considered. However during 1978, development trials were conducted and while
we could not produce a coin of the same hardness as a gold-copper alloy, we could
certainly produce a good coin with reasonable hardness, considering that the coin
would not be in general circulation and subject to "pocket" abuse.

HARDNESS

As mentioned, there was concern about the ability of the blank supplier to not
only produce the required hardness, but to maintain the required tolerance.

In addition, the hardness of the resulting coin was important so that it could be
handled without becoming badly marked even though it would not be in general
circulation.

A number of preliminary tests were undertaken by the Mint to establish that the
hardness specification was attainable and that the resulting coin would meet our
requirements.

However, the only really true trial was the fabrication of thousands of blanks and
coins before concluding that the hardness specification was valid.

The initial production run of blanks demonstrated that it was difficult to obtain the specified hardness values of 43 ± 3 RB. However following development work with the blank supplier, very consistent hardness readings were obtained. Our specification was modified from the 43 ± 3 RB for a single reading to an average value of no less than 38 RB for five readings. In addition, the spread between the high and low value was established at a maximum of 6 RB hardness points. By controlling the spread to 6 points between a high and a low value and accepting 38 RB as the minimum average value, all blanks produced met the revised specification. Hardness values for lots tend to average 40 to 42 RB.

This was not only successful with the blanks received during 1979, but has continued to be so during 1980, even though we are using a different blank supplier.

COIN DESIGN

Taking into account that this bullion coin was a first for Canada, the design on both the obverse and the reverse was of extreme importance.

As has been mentioned previously, it was a Marketing advantage for a coin which was to have world distribution to be a monetary unit of Canada. To provide this, a $50 face value was recommended by the Department of Finance.

Since it was now a coin of the realm, it would carry the standard effigy of the Queen, which is present on all coins that the Royal Canadian Mint strikes for Canada. However, the reverse design was open for discussion and it was felt within the Mint and with those with whom we talked outside the Mint that the reverse of the coin should have a very distinctive design and one that would be symbolic of Canada.

As you all very well know, the symbol of Canada which is most recognized throughout the world is our flag and the Maple Leaf thereon. Therefore, a number of Maple Leaf designs were prepared which included a cluster of maple leaves and a single Maple Leaf.

Some of the designs were taken from photographs, while some were taken from samples of real leaves. After reviewing the various designs, it was decided that a single maple leaf was the most appropriate and this idea was fully developed. The single leaf design along with the recommended lettering was approved in February 1979.

The pencil drawing, made by our Engraving Department, was very detailed and included all of the leaf detail in terms of veins and relief and highlights, and as we all know from the final product, the leaf does look real.

Following the program approval in February of 1979, our Engraving Department produced the required models and Master Tooling.

In this particular instance, they were asked very specifically to produce a maple leaf in the model form that would result in a coin with a leaf that was so much like a real maple leaf that one would have to conclude when one looked at the coin that a Gold Maple Leaf had fallen from the tree and was resting there.

I think for all of those who have seen the design and the actual struck coin, they would have to agree that the Maple Leaf does truly look alive.

SIZE

There were a number of technical details to be considered when designing a coin. Among them being:

(a) should it be comparable in size and diameter to the Krugerrand,
(b) if pure gold was to be used, should the diameter be reduced so as to produce a thicker coin and hence create an image of greater value, or
(c) should the diameter be larger than comparable coins and hence thinner?

In reviewing the above options as well as (a) packaging requirements (b) press capacity (c) marketing impression (i.e. balance between diameter, thickness and weight), it was decided that a coin of approximately 30 mm diameter would be the best. To arrive at 30 mm, we had considered diameters such as 29, 30, 31 and 32 mm. However, it was obvious that there was really no real difference between 29, 30 or 31 and therefore, 30 was chosen.

The final specification was 30 mm diameter with a nominal struck edge of 2.8 mm.

It should be mentioned that we had had considerable striking experience and very good success in both the market place and in the striking of a 27 mm half ounce gold coin and the increase to 30 mm and a one ounce weight was in our estimation, a very logical step to take and one in which we feel has resulted in a very attractive coin.

MINTING CONSIDERATIONS

When examining a program with the importance of the Maple Leaf, it was necessary to assure that the alloys, the design, the potential striking and the entire process of how this coin was to be made were carefully considered. One of the principle factors for choosing a pure gold coin in addition to its yellow gold colour was one of ease of processing. A pure gold coin of 999 purity could be easily made if the starting material was a 9999 gold. In addition if starting with 9999 pure gold, there would not be any need for refining the scrap. This point was discussed very thoroughly with the potential blank fabricators. They felt they could recycle the scrap up to three times before the gold would have to be refined 9999 purity.

By reducing the need for re-refining or re-alloying, we would also reduce the cost of blanks to the Royal Canadian Mint.

Therefore during the 1979 program in which 1 million coins were produced, the blank fabricator, to our knowledge, did not have to refine any of the scrap gold from his process and certainly all of the coins that we would have had to scrap for various reasons, were melted down into a pure bar and sent back directly into the process without the necessity of re-refining. This reduced need for re-refining has certainly resulted in an energy saving, as well as costs and still provided a guaranteed purity in the finished product.

Following our success with the 1979 program and the fact that random samples of coins chosen for assay were found to be 9999 pure, it was decided to increase the purity requirements for the 1980 coin from 999.0 minimum to 999.8 minimum. Random sampling of blanks for 1980 has supported the fact that all 1980 coins are 999.8 minimum purity.

COST

Therefore, as a result of the decision to have a pure gold coin without alloying elements and ease of scrap recycling, the precious metal blanks were obtained at a very attractive price.

In addition, we have found the striking of the coin to be very economical with very low percentage of rejects. In fact, there is less than 1% rejects for all reasons including over/under weight.

The coin blanks are automatically fed, single struck and visually examined at a rate of 10,000 to 12,000 per 8-hour shift.

FINAL SPECIFICATION AND LAW

Following this development work, both the design and technical specifications were written into law.

CHANGES IN 1980 FROM 1979

As mentioned previously, following the good experience with the blank manufacturer in starting with 999.9 gold, we produced a coin that was consistently 999.9 pure. Therefore, it was decided to propose a change to the law in 1980 so that we could take advantage of the fact that we were using 999.9 gold and not 999.0. A change was made in the law to not only change the purity from 999 minimum to 999.8 minimum, but also to change the nominal weight of the coin from 31.185 grams to 31.150 grams and at the same time, to decrease the weight tolerance specification from .05 grams to .04 grams.

These two changes reduced the amount of gold that must be given away in each coin so as to guarantee one ounce of fine gold in each coin sold.

Since we all can appreciate that the price of gold has accelerated rather dramatically, this reduction in gold give-away has resulted in reduced operating costs for 1980.

We are pleased to report the changes that have been made in 1980 to specifications have resulted in a net savings to the Mint operation, which in turn is passed on as a savings to the Canadian Government.

PACKAGE DESIGN

In launching a program such as a gold bullion coin, it was necessary to provide a package which would satisfy a number of criteria among which were:

    a)   a pleasing appearance
    b)   security in transit
    c)   ease of handling for the banks and distributors
    d)   a pleasing appearance to that purchaser who may buy ten coins.

In addition to the above, we had to make a choice of how many coins would be packed in an individual container as well as a master pack.

Packaging coins into multiples of 5, 10 or 20, all would be convenient if used

alone, whereas a combination would create inventory problems and additional costs
for packaging material.

Therefore, after reviewing many layouts and carefully considering whether 5, 10
or 20 should be the sub-unit, we arrived at what we felt was the optimum design.
We chose ten as a multiple and designed a translucent white plastic tube, similar
to the one used for the Krugerrand, that would exactly hold 10 coins. After
deciding on a total of 500 in a box in multiples of 10, a layout was developed
which would ensure that the package:

> a) would provide support in transit (i.e. keep all coins
>   from rattling or rubbing against one another);
> b) permit tubes to be easily removed;
> c) be stackable;
> d) permit steel strapping and seals to be affixed;
> e) be easily identified as a Royal Canadian Mint package.

The layout chosen provides four across and a combination of three and three, along
the length of the box, stacked two high, to give a total of 480. At the centre
of the box, a filler piece was added which contained two tubes of 10 coins. This
filler piece is designed so that the two centre tubes of 10 can be easily lifted
and permit ease of removal for balance of tubes.

The Master Pack was also designed so that the metal strapping when applied could
not be removed by sliding it off the box. It can only be removed by cutting and
thereby destroying the seal.

The bottom and top of the box were also designed to permit nesting and hence they
can be stacked three, four or five high with little risk of the boxes toppling.
I have a sample of the master pack which will more clearly show these design
details.

The boxes, after being strapped with metal, are sealed with a Mint seal. Strapping is done in both directions.

The finished package then contains 50 tubes of 10 coins, metal strapped in two
directions, sealed with the Mint seal and ready for the application of the
customer's name and address.

Within the Mint, we have found filling of tubes, packing in boxes and final strapping operation to be both efficient and effective.

We feel that the time and effort spent on the design considerations for this total
package have resulted in a very economical, attractive and secure package.

OPERATIONAL HIGHLIGHTS

No discussion would be complete without a review of the various production operations required to produce the finished coin. The blanks arrive at the Mint in
boxes of 960 with ten boxes to a lot. Based on these incoming lots of 9600 blanks,
samples are taken for the following quality checks:

(a)   20 samples for assay to assure that the gold content is 999.8 or better;
(b)   200 samples on which five hardness readings are taken;
(c)   all the blanks are weighed in sub-lots of 240 to assure that on average,
       each of the incoming blanks meets the average weight criteria; and that the
       gold received can be reconciled with gold purchased for the program;

(d)   the 200 samples for hardness determination are also visually inspected to assure that they are free of major surface imperfections or contaminants.

Following the acceptance of the 9600 blank lots, each 24 blanks are placed in a tray and 10 trays are combined in a specially-made box.

When blanks are required for the press room, they are delivered to the press room in these boxes. Only 960 blanks and/or coins may be in the press room at any one time and usually as soon as 480 of these blanks have been struck as coin, the coin is returned to the vault and an additional 480 blanks delivered. All vault/press room transactions require signatures for the blanks or coins.

In this manner, very tight control can be maintained and at the same time, there is no interruption in the striking and 10,000 to 12,000 blanks can be struck as coin in a normal 8-hour shift.

Following the striking operation, the coins are returned to the vault where each coin is individually weighed.

We have purchased for this operation a Mettler balance with mini-computer which signals over/under weight from a predetermined standard as well as prints the individual weights, and finally for each 250 produced, produces a histogram so that the statistical information is readily available.

The use of this computer has resulted in extremely good weight control and at the same time, it has reduced the operating cost as we do not have a number of clerks calculating average values nor recording the values. These are recorded automatically on the scale output equipment. To provide a verification of the scale, standard weights were produced for minimum/average/maximum weight. These weights have been certified. After each weighing of 500, the balance is checked with these standard weights to ensure correctness of weight. Following the weighing operation, the coins which are kept in lots of 250 are delivered to the packaging area where the coins are placed in the plastic tube and then the top applied and firmly attached by use of a small pneumatic ram. As soon as the top is on, the plastic tube of ten coins is taken by another operator who ensures that:

(a)   they do not rattle;
(b)   the top is on correctly and that there are not too many coins, and;
(c)   the weight of the tube is within the tolerance for ten coins.

Following this, the coins are placed in the master pack, a sample of which I have in front of me. The master pack is then strapped in the two directions and the seals applied automatically.

The master pack is weighed so as to ensure that weight is within tolerance for 500 coins.

After strapping and sealing, the coins are returned to the vault for intermediate storage until shipping instructions are received.

Signatures are required from all personnel who handle bullion coins, no matter what the operation. This has resulted in very fast identification of any miscounts or "lost" blanks and to date, there have been no problems which could not be resolved within 15 minutes.

Proof Coining of Silver

Joseph Kozol
Director of Technical Services

The Franklin Mint, Inc.
Franklin Center, PA., U.S.A.

ABSTRACT

The attractiveness of sterling silver has made it a desirable alloy for flatware, hollow ware, jewelry and coinage for many years. Very little information has been published, however, on the means by which proof quality coins are made. Starting with the materials used in melt makeup, the processes by which silver is melted, cast, rolled, annealed, and fabricated into proof medals are discussed. At The Franklin Mint, silver is continuously cast into bars, cold rolled to size and processed into planchets, taking great care in each step to maintain cleanliness and to minimize material defects. Handling must be done so as to avoid stains, nicks, dents or scratches in all phases of blank preparation. Racks and fixtures are custom designed to the same purpose. Proof coins are struck between highly polished, clean dies with frosted images and mirror finish background.

KEYWORDS

Proof coining; proof medals; sterling silver medals; continuous casting; cold rolling of silver.

INTRODUCTION

This paper describes the manufacture of proof coins and medals at The Franklin Mint, Inc. The term "coin" applies to all coin-of-the-realm (currency) but is also used to describe certain small size non-monetary pieces; e.g., Mini-Coin. A coin turn consists of a reverse design which is inverted to that of the obverse. The term "medal" is used where the emphasis is on artistic or commemorative value. For a medal turn, the reverse design is right side up in relation to the obverse. At The Franklin Mint, a proof coin or medal represents the ultimate in quality and is a culmination of the efforts of artists, sculptors, engravers, minters, and numerous other personnel. A proof finish consists of a finely frosted image raised in relief against a mirror finish background. Quality is, of course, subjective but the finished product is required to be free of dirt, stains, scratches, extruded fins and other visible defects.

Brief consideration will be given to the physical metallurgy of silver base alloys, followed by a description of the processes involved in the manufacture of proof coins and medals.

Metallurgy of Silver-Base Alloys

The silver-copper binary system is a classic eutectic system, as shown in Fig. 1.[1]

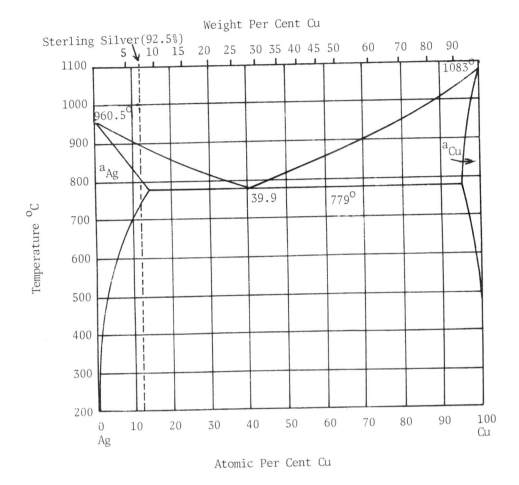

Fig. 1. Ag - Cu phase diagram

The Franklin Mint coins alloys in this system from pure (999 fine) silver down to
500 fine, although this paper deals with the coining of 999 fine and sterling
(92.5%) silver only. The vertical line representing 92.5% silver in the equilibrium diagram shows that sterling silver passes from a single-phase to a two-phase
region on cooling, indicating that sterling silver is a precipitation hardenable
alloy. It is generally overaged on annealing to provide an optimum hardness level
for coining. If the alloy were rapidly quenched from the annealing temperature,
the silver rich single phase would be retained in a supersaturated solid solution
and the alloy would remain soft. With very slow cooling, the copper rich phase
precipitates, the alloy eventually overages and is again soft. Intermediate (e.g.,
air) cooling from the annealing temperature provides a visible copper rich

[1]Hansen, M. (1958). Constitution of Binary Alloys. McGraw-Hill Book Company, Inc.,
    New York. pp. 18-20.

precipitate and, with proper adjustment of parameters, a hardness optimum for coining. Because the copper rich phase oxidizes more readily, the single-phase fine
silver is more tarnish resistant than the two-phase sterling silver alloy.

PREPARATION OF SILVER ALLOYS FOR PROOF COINING

Continuous Casting

Starting materials in melting are 999 silver bullion and (for sterling) electrolytic (ETP) copper and copper phosphide as a deoxidant. The amount of phosphorus
is controlled carefully in a narrow range. Too little phosphorus results in
blistering on annealing; too much phosphorus causes a pink appearance and can
result in casting problems. Clean scrap from in-house reject medals and blanking
web scrap are also utilized to a large extent. Five thousand troy ounce charges
are made up and loaded into an induction melting furnace under a reducing flame.
A charcoal cover is used to retard oxidation. After melting, the alloy is poured
into an electric holding/casting furnace and thoroughly stirred.

At this point a sample of each charge is taken for assay and the charge is identified so that all cast bars can be traced. A dip sample is beaded by pouring it
into water and sent to the Assay Lab. The Lab utilizes a Gay-Lussac method with
potentiometric determination of end point. The titrimeter is standardized twice
daily with a silver standard of 99999 purity. Cross reference checks are made
periodically against outside Laboratories. Assay approval is required prior to
rolling.

The continuous casting process is illustrated in Fig. 2. Molten silver is poured
through a slot in the bottom of the crucible through a water cooled die. The bar
solidifies in the die and is held by pull rolls as it exits with the speed of casting controlled at approximately 10 min/6 ft. bar length to keep the solidification
zone at the proper level. If the liquid zone drops too low, a "breakout" occurs
and the metal pours out. This restricts the speed of casting. A specially designed
copper cooling jacket was developed at The Franklin Mint to extract heat from the
die fast enough to enable increased casting speed without breakouts and still maintain a desirable as-cast structure. At a predetermined bar length (6 ft.), a flying
saw automatically cuts the bar. Bars are cast from 1/2 inch to 1 inch thick by
5 1/2 inches wide. Continuous casting produces sound, clean alloys for proof
coining.

Surface Finishing & Rolling

While production is a continuous process, casting is not truly continuous in the
sense that the bar starts and stops. Segregation of copper, which occurs during
movement of the bar through the die, causes the formation of curved "tiger stripes"
at the solid liquid interfaces. To eliminate this surface discoloration and other
potential surface defects, the bars are scalped on a horizontal milling machine.
This produces a clean surface, free of abrasive particles and other contaminants
for rolling to final size.

After assay approval, the silver bars are cold rolled. Sterling bars are rolled to
approximately 0.150 - 0.180 inches thick on a 2-high breakdown mill and then
finished to final size on another 2-high mill. A small 4-high mill is used to roll
to thicknesses less than about 0.035 inches. While sterling bars are rolled
directly to size, the large recrystallized grain size of fine (999) silver results
in an "orange peel" effect if the bars are not intermediate annealed. Fine silver
receives about 70-80% reduction in thickness and is then intermediate annealed

Joseph Kozol

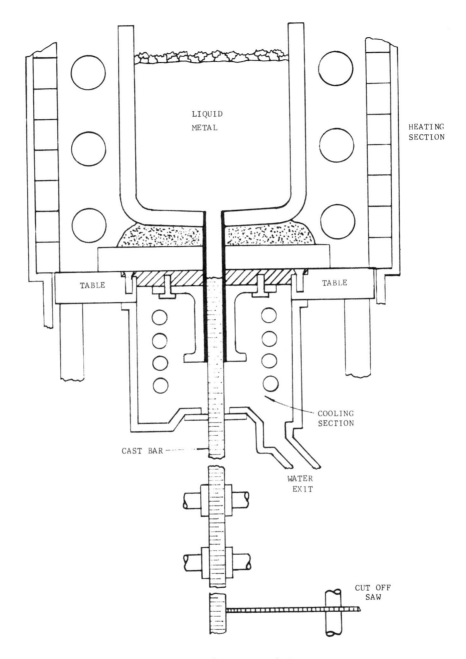

Fig. 2. Continuous casting apparatus

before final rolling. The rolls of both the breakdown and finishing mills are polished with diamond lapping compounds at the beginning of each shift and when there is a change in alloy, to produce a mirror finish on the strip. Transfer tables are utilized to minimize handling of the strip between operations and these are also cleaned daily.

Blanking

After the strip is rolled to size, planchets are punched out on mechanical presses without lubrication, usually on multiple punch die sets. The optimum type of edge for proof coining is approximately 50% fracture surface and 50% shear, with a minimal burr. As blanking continues and the punch dulls, the burr increases. Because there are no subsequent metal removal processes, it is necessary to keep the burr size to a minimum. This is accomplished by keeping the punch and die sharp and maintaining proper clearances between them. Clearance is normally 3-4% of thickness on the radius. For special applications, fine blanking is performed with lubrication; clearances are tighter and the edge is essentially 100% shear surface. From the time the blanks are punched, they are handled with vinyl gloves, with great care taken to avoid stains, nicks, dents, or scratches.

Annealing & Edge Rolling

Sterling silver blanks are in a hardness range of $R_B 80-90$ after undergoing in excess of 70% reduction in thickness in cold rolling. In order to soften them for coining, they are annealed at 1250°F to 1300°F in a continuous belt furnace, in an exothermic atmosphere at an air-to-gas ratio of about 7.5:1. Hydrogen content is approximately 5-6% and carbon monoxide is about 6-7%, providing an economical reducing atmosphere for bright annealing. The blanks do emerge with a slight haze and must be subsequently brightened. Optimum hardness for coining is approximately $R_B 20$ for sterling silver. After annealing, the blank rim is upset in an edge rolling operation, in order to smooth out the fracture surface, to provide proper diameter for clearance in coining and to make the rim easier to fill. Work hardening of the rim helps to avoid formation of a fin on coining; i.e., the movement of metal will fill the die prior to extrusion of a fin at the hardened rim.

Burnishing

Depending on their geometry, blanks are burnished in either rotating barrels or vibratory tubs, with 5/32" spherical stainless steel shot and alkaline cleaning compounds. There is no metal removal in these operations, only planishing, which produces a high luster. The stainless steel media must be maintained to avoid contamination and the formation of irregular shapes, which would not produce a bright surface.

The rotation of the barrels produces a sliding, scrubbing action as the shot "landslides" across the blanks. The barrels are lined with neoprene to prevent contamination. The ratio of media to blanks is maintained at 8:1 or greater to avoid blank-on-blank impingement. The cleaning compound provides lubricity and prevents staining and tarnish; the blanks must be thoroughly rinsed and dried quickly to avoid staining. The Franklin Mint utilizes deionized water rinses to prevent mineral stains.

In vibratory burnishing, eccentric weights are used to develop a vibratory motion which causes the media and work to progress around the tub in a helical motion. Specially designed fixtures are utilized to hold heavy parts or those that are irregular in shape.

After rinsing in deionized water, barrel burnished and vibratory burnished blanks
are dried in rotating barrels with corn cob fines. The cob must be kept free of
moisture to avoid staining.

Coining

Although there is generally no heavy dirt or oil present, blanks are given a final
aqueous cleaning prior to coining, to remove any traces of grease, dirt, corn cob
or tarnish. They are first dipped in a neutral detergent, followed by a double
counterflowing rinse. The blanks are then dipped in a tartaric acid solution,
thoroughly rinsed and dried in Freon to avoid water staining.

Coining is performed in a laminar flow clean room atmosphere in which the incoming
air is filtered to remove dust and lint particles. All employees in the room are
required to wear lint free-static free caps and coats. Small particles, if embedded
in coining, cause visible rejects. Other types of rejects are caused by scratches,
droplets of oil or other stains, or extruded fins. The scrap rate in proof coining
may be as low as 10% or as high as 50% of the output. The larger the medal and
the more open, polished area, the greater the likelihood of a visible defect causing
a reject. Extra care in handling is required all through the process as the size
of proof medals is increased.

The blanks are hand fed by the operator and double or triple struck in a closed die
coining operation; i.e., they are restrained by a collar (which forms the edge)
between highly polished dies with frosted images and a mirror finish background.
Dies must be cleaned often during the operation, generally with alcohol on soft
polishing cloths.

The relatively low yield strength and low rate of work hardening of both sterling
and fine silver provide excellent coinability. As described earlier, a hardness
level of R_B20 is considered optimum for sterling silver. At this hardness a high
relief (approximately .015 - .020" for a 1 1/2" medal) can be filled at a tonnage
level which will not drastically lower the life of the dies. Coining is performed
at loads up to 120 tons/square inch. As tonnage increases, in order to obtain high
relief, die life is reduced proportionately.

There are, of course, other factors affecting die life, which will be discussed
briefly. Dies for striking proof coins are made of low alloy tool steels and are,
in most cases, consumable vacuum melted for optimum cleanliness. Although the
artistic features are of paramount importance, it is necessary to consider the
effect of stress raisers for extended die life. Sharp angles and designs which
would create stress concentrations in a relatively brittle tool steel must be
avoided. The engraver, in translating the sculptor's work into medallic design,
must take this into account and maximize radii wherever possible. The location and
amount of high relief areas in the art work is an important factor in determining
the final tonnage required to fill the die and hence can have a significant effect
on die life. For this reason, art work is generally controlled so that a high
relief area on an obverse die is compensated by a low relief in the corresponding
area of the reverse die. In manufacturing dies. grinding of hardened tool steels
must be performed so as to avoid overheating which could over temper or, in extreme
cases, induce hairline cracks which would drastically lower die life. Correct
alignment in the press is also critical in obtaining uniform pressure during
coining. Under optimum conditions, approximately 10,000 (double struck, 1 1/2"
diameter) sterling silver proof medals can be produced from a single set of dies,
providing excessive tonnage is not required.

SUMMARY

As described earlier, sterling and fine silver are subject to tarnishing on exposure to the atmosphere. In order to preserve the beauty of the proof medals, they are sprayed with a fine coating of clear lacquer prior to packaging. The lacquer contains tarnish retardants and ultra violet stabilizers to prevent yellowing. In summary, considerable effort and attention to detail is expended at each stage of manufacture, in order to assure that the discerning collector will enjoy the lasting beauty of silver proof medals.

Proof Coining of Karat Gold and Platinum

J.W. Simpson

Franklin Mint, Franklin Center, PA.

ABSTRACT

Commonly thought of as only used for jewelry, the karat golds (22, 18, 14, and 12) and platinum are regularly processed to make proof quality coins and medals at The Franklin Mint. Proof coins are defined as having a frosted image surrounded by a bright, mirror-like background. For gold and platinum the alloy selection criteria of precious metals content, color, grain size control, and legal requirements are discussed. Following this, the processing requirements for proof coining are presented. Special procedures used for casting, rolling, blanking, and burnishing of gold alloys are discussed. Processing of purchased platinum planchets is also included.

KEY WORDS

Coins, Gold, Medals, Platinum, Proof

INTRODUCTION

The Franklin Mint is a private organization engaged in the production of high quality proof coins and medals using a wide variety of materials. One segment of the operation involves the use of 22, 18, 14 and 12 karat gold and platinum to make medallions for pendants, commemorative medals and coinage for other countries. Table 1 lists and identifies the major alloys now in use. Alloys used for foreign currency are identified with the fineness system while alloys for medallic applications are referred to by the karat system.

This article describes how these alloys were selected and how they are processed to arrive at a proof quality product. The information presented here has been obtained mostly in response to specific requirements and was not the result of extensive research into the gold and platinum alloy systems. Therefore, it should not be assumed that the alloys presented here are the only ones available. For an excellent metallurgical analysis of karat golds the recent article by McDonald and Sistare (1978) is strongly recommended.

Before discussing alloys, a brief definition of proof quality is required. To be labeled proof, a coin or medal must contain a frosted or matte image area surrounded by a polished, mirror-like background. Beyond this there are no universally accepted quality standards. For a coin to be called proof at The Franklin Mint, the

TABLE 1 Alloy Compositions

Alloy	Au	Ag	Cu	Zn	Pt	Pd
Gold						
900 fine (22 kt.)	90	6	4			
18 kt. yellow	75	15	10			
18 kt. green	75	22	3			
14 kt.	58	4	31	7		
500 fine (12 kt.)	50	5	37	8		
Platinum					99.9	
Platinum-Palladium					98.2	1.8

polished area must be free of all visible defects, including dirt, lint, scratches and unevenness resulting from the grain structure of the metal being coined. To obtain such a finish, two basic ingredients are needed; clean, bright blanks or planchets and dies containing the polished and frosted areas. Nothing is done to the product after coining. Anything coming between the die and planchet will result in rejection of the coin. Therefore, processing of the alloy into planchets must not introduce any mechanical damage or contamination that cannot be removed prior to coining.

ALLOY SELECTION

Precious Metal Content

At The Franklin Mint, Marketing plays a very active role in determining the karat gold selected for a specific job. The interplay between selling price and perceived value must be analyzed for each product. Whether the product is to be a pendant, commemorative medal or foreign coinage, all effects the selection procedure.

However, engineering considerations also strongly effect the decision. For a given diameter coin, the gold content can be reduced by making the coin thinner and by using a lower karat gold. Unfortunately, these two factors are interrelated. The minimum coin thickness is determined by the ability to successfully coin a planchet without exceeding the yield strength of the die. Soft alloys can thus be coined at lower thicknesses than harder ones. Table 2 lists the minimum coining thickness for a one inch diameter yellow gold coin with artwork about .010 inches (.25mm) high The thickness increases as gold percentage is lowered.

TABLE 2 Minimum Gold Planchet Thickness for One Inch Diameter Coin

Alloy	Min. Thickness (in.)	Gold Content (tr. oz.)
900 fine	.035	.26
18 kt. yellow	.055	.35
18 kt. green	.035	.23
14 kt.	.050	.26
500 fine	.040	.20

By obtaining similar data for other coin diameters it is possible to plot the gold content against coin diameter for the alloys listed in Table 1. This has been done in Fig. 1. It is evident that the 18 kt. yellow and the 14 kt. alloys are not cost effective because a 900 fine coin (22 kt.) actually contains less gold. Significant savings are not realized until a 50% gold level is reached. Therefore, the two most commonly used gold alloys at Franklin Mint are 900 and 500 fine (12 kt.). In fact, for foreign currency applications these are the only karat gold alloys specified.

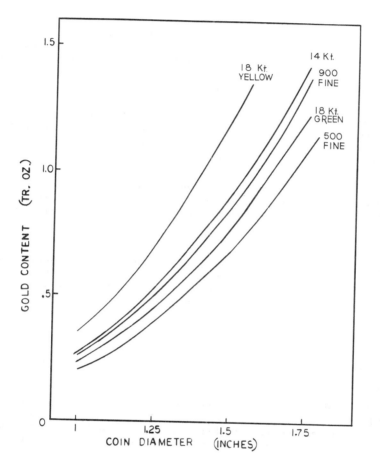

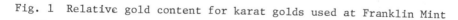

Fig. 1 Relative gold content for karat golds used at Franklin Mint

An interesting effect occurs in the 18 kt. gold alloys. Because of a complex solid state reaction, gold alloys containing 75% gold are made very hard by the addition of copper. As will be discussed later, copper is necessary to balance the effect of other alloy additions to create the yellow-orange color of gold. Therefore, it is not possible to have a yellow 18 kt. alloy that is soft enough to coin economically.

To resolve this dilemma, the color requirement is dropped and silver is added to the alloy to replace the copper. The resultant alloy listed in Table 1 as 18 kt. green is comparable in properties to the 900 fine alloy. However, the greenish color must be covered by gold electroplating to create a pleasing product for use in pendants and commemorative medals. The electroplating precludes use of the alloy for foreign currency applications.

For platinum, alloys must contain at least 95% platinum to be so labeled. Table 1 lists the two platinum alloys currently in use at Franklin Mint. The pure platinum is used occasionally for commemorative medal application while the 1.8% palladium alloy was developed for the Republic of Panama. Lowering the precious metal content was not a factor in these alloys as will be discussed later.

Color

Once a specific karat level is selected, a suitable color must be obtained. The color of gold alloys is determined by the ratio of alloying elements. Silver gives gold a greenish tint while copper moves gold towards a red color. Zinc is neutral and is used essentially to dilute the effects of silver and copper. Figure 2 shows a qualitative color diagram for the Au-Ag-Cu system (Leuser, 1949).

To produce white golds, nickel is commonly added. However, this produces alloys that are too hard for coining. By substituting palladium for the nickel, a coinable white gold can be produced. The relative high cost of palladium has prevented further use of this alloy at Franklin Mint.

Other alloying elements such as tin and indium have been briefly tested as replacements for copper, but either severely harden the alloy or produce irreversible color effects. Surface altering techniques have also been tested with little success.

Grain Size Control

It is almost impossible to process commercially pure platinum without causing excessive grain growth that could interfere with proof coining. For karat golds, grain growth is generally not a problem. Grain growth in platinum was solved by adding 1.8% palladium. Nickel was also successfully tested. However, for the reason described in the next section, the palladium alloy was selected.

Legal Requirements

Often alloys are determined by legal requirements within individual countries. For example, a platinum alloy containing 5% palladium was recommended to Panama for the 200 Balboa Coin. As mentioned, previously this alloy satisfied the hallmarking guidelines and could legitimately be called platinum. By adding the full 5% palladium, a modest cost savings could be realized. However, the Panama Government selected and legislated the 1.8% palladium alloy listed in Table 1. For gold alloys France and Sweden will not accept jewelry items such as coined pendants that are less than 18 kt. gold.

PROCESSING FOR PROOF COINING

The Franklin Mint melts, casts, rolls, blanks and burnishes all the gold alloys previously discussed. Because of the high melting point of the platinum alloy, annealed blanks are purchased. The next sections discuss processing of gold alloys only. A section on platinum is included at the end.

Melting and Casting

All melting is done by induction. Charges range in weight from 500-700 tr. oz. depending on the alloy. Bar size is approximately 1.2 in. x 5.5 in. x 11 in. Extensive planning is done to ensure the most complete use of all starting bullion. For example, the 14 kt. and 12 kt. alloys have the same Ag: Cu: Zn ratio so that scrap from one can easily be used for the other. The 18 kt. green alloy is made by adding silver to 900 fine scrap. Although providing for efficient use of gold, such a system can lead to oxygen pick up because of the many remelt cycles that

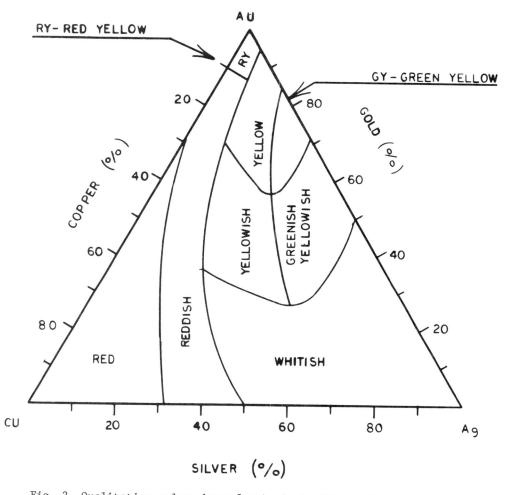

Fig. 2   Qualitative color chart for Au–Ag–Cu Alloys (Leuser, 1949)

often occur. Proprietary techniques for solving this problem have been developed. The 14 kt. and 12 kt. alloys have the further problem that a loss of the highly volatile zinc will result in a color change for the alloy.

After casting, all surfaces of the bar are machined to remove the solidification pipe and surface oxidation. The resultant scrap is immediately recycled. Each bar is assayed and approval is required before proceeding.

Rolling and Blanking

Rolling is done cold using separate, two-high, breakdown and finishing mills with polished rolls. To ensure smooth clean strip, the rolls are polished at the beginning of each shift with diamond compound. The use of highly polished rolls is a very important first step in the production of proof coins. Lead-in and run-out tables are also cleaned daily to remove dirt that may transfer to the strip.

The rolling schedule varies greatly depending on the alloy and final gauge. Very little 18 kt. yellow gold is processed because it challenges the capacity of the mill. The 14 kt. alloy is the hardest of all the popular alloys and often requires three intermediate anneals to reach final thickness. The 900 fine and 18 kt. green alloys can be rolled directly to gauge but are often given an intermediate anneal for grain size control.

Blanking is done on multiple punch progressive die sets to optimize the yield from each strip. Blank yields as high as 65% are often obtained. The resultant web is immediately sent back for remelting.

Annealing

All annealing is done on traveling belt furnaces with an exothermic atmosphere made from partially burned natural gas. The alloys containing zinc must be covered to prevent volatilization of the zinc from the top surface. This would result in a color change for the surface that looses zinc. For the other alloys care must be taken to prevent the excessive grain growth that can occur when the temperature is too high.

Rimming and Burnishing

After annealing, the planchet edges are rolled to assist in properly filling the edge of the coin. By this time, the planchet has received some handling damage and undoubtedly picked up some dirt. To remove this and prepare for coining, the planchet is burnished by tumbling it with steel shot and soap. Although the blank has a peened appearance, it is very clean, bright and smooth enough for coining. The slight roughness imported by burnishing is easily coined flat. Vibratory burnishers are also available for larger planchets, but are not generally used for gold. At this point, all planchets are individually weighed to assure compliance with minimum guaranteed weights.

Coining

Immediately prior to coining, the planchets are given a final solvent cleaning to remove any traces of burnishing compound and dirt. They are then hand-fed into the coining press. Even with all these precautions and cleaning operations, rejects at coining generally run 15-20% and for coins with large polished areas can run as high as 30%. Reject rates higher than 30% generally indicates that one of the cleaning steps was not properly executed.

The total yield of acceptable proof coins is generally about 30% of the initial cast weight. Recyclable scrap is generated from machining of the cast bar, web scrap from blanking, underweight planchets and coining rejects.

Platinum Processing

Purchased platinum blanks are weighed, rimmed and burnished at The Franklin Mint. Material problems that have occurred are porosity and excessive grain size. These can be catastrophic since it is generally impossible to detect them prior to coining. Even if they could be detected, nothing can be done at Franklin Mint to allow successful proof coining. Both the porosity and excessive grain size were minimized when the 1.8% palladium alloy was selected.

SUMMARY

Gold and platinum alloys are also widely used for aerospace, electronics, dentistry and other industrial applications. Each industry has its own alloy selection

criteria based on the eventual use of the product. The purpose of this article
has been to qualitatively discuss the criteria used to select alloys for proof
coining and to briefly present the special processing procedures involved in trans-
forming the alloys into finished proof coins.

REFERENCES

Leuser, J. (1949). Metall, 3, 105-110, 128
McDonald, A.S. and G.H. Sistare (1978). The Metallurgy of Some Carat Gold Jew-
    ellery Alloys. Gold Bulletin, V. II, no. 3, 66-73

New Developments

New Developments in the Use of Gold in Electronics and Communications

Robert R. Davies

The Gold Institute/l'Institut de l'Or
1001 Connecticut Avenue, N. W.
Washington, D. C. 20036

ABSTRACT

The reliability of gold conductors makes them highly desirable for a variety of applications in electronics and communication systems. Refinements, however, continue to be made in these applications which permit the gold to be used more effectively. Both state-of-the-art technology and recent improvements are reported.

Among the improvements are the use of iron and cadmium rather than cobalt and nickel as hardening agents for plated golds. New selective plating processes rapidly deposit gold in very well defined areas and to very well defined thicknesses. A unique alloy of gold with indium catalyzes the formation of a polymer film which makes the gold an excellent sliding conductor. Gold is used in a variety of ways in the growing technology of hybrid microcircuits. Wear resistant gold inserts are used for television channel selectors and other growing applications.

KEYWORDS

Hard golds; gold wire; gold thick films; contacts; selective plating; oxidation resistance; microcircuits.

VERSATILITY

Gold is a very versatile metal whose special properties make it uniquely suitable for a number of applications in the fields of electronic and communication engineering. The Gold Institute/l'Institut de l'Or continually monitors the development of these applications, some of which are quite commonplace, and some of which are in the forefront of new high technology. Gold is in our television sets, in our pocket calculators, and in the increasingly popular electronic games played by children and adults alike. It is performing an essential role in the electronic brains of the new generation of household appliances. Attention has been drawn to these uses in recent issues of The Gold Institute's bimonthly publication entitled "The Gold News/Nouvelles de l'Or". In the past the same publication has also carried reports on gold plated telecommunication antennae and on the use of gold mesh to shield the communications systems of commercial aircraft from electromagnetic interference.

Frequently it is gold's excellent electrical conductivity which is utilized.
Gold, however, has many other special properties. For certain applications it is
desirable to use gold's superb ability to reflect infrared radiation. This particular attribute of gold is, for example, effectively utilized in improved photo-
copying machines. In such machines a toner is electrostatically deposited on the
copy paper. The paper subsequently passes between two smooth cast aluminum surfaces plated with pure gold. One of the gold plated surfaces is flat while the
second surface is parabolic. Both gold plated surfaces reflect intense infrared
heat radiation from a quartz lamp located between them. This radiation causes the
plastic powder carrying the carbon black in the toner to melt so that the carbon
black diffuses into the paper and instantly forms the desired permanent image.

Often it is a combination of characteristics which make gold the preferred material. In hybrid microcircuitry it is gold's bonding behavior as well as its electrical conductivity which makes it desirable to use gold coated wire to join semiconductor chips. Excellent thermocompression bonds can be formed between gold
coated copper lead wires and gold coated lead frames.

The copper wires may measure between 0.2 and 10 mils (or thousandths of an inch)
in diameter. These wires are typically plated first with nickel and then with
pure gold deposits measuring 50 microinches (about 1-1/4 microns) in thickness.
Very fine wire made of pure solid gold is also used extensively in the fabrication
and connecting of hybrid microcircuits.

Other types of gold coated wire are used in the vacuum tube industry. Molybdenum
and tungsten wires coated with hardened gold can withstand the very high temperatures encountered inside the tubes. The gold is generally hardened by the inclusion of 0.25% of either cobalt or nickel.

RELIABILITY

An attribute which generally ranks very high for recommending gold over other materials is gold's reliability. This reliability is largely due to gold's superior
resistance to oxidation and tarnish. A gold part may initially cost a few cents
more than a copper part, or one made of silver. Many dollars, on the other hand,
would be sacrificed in terms of lost operating time and repair expenses if the
electrical properties of the non-gold substitute metal would deteriorate due to
the formation of oxide or sulfide films. Such films do not form on gold.

The freedom from oxidation and tarnish films is of particular importance for certain types of make-break contacts. The contacts used in telephone switching systems, for example, must perform reliably time after time. Silver contacts are
ideal for large current and large voltage devices such as circuit breakers and the
contacts which start and stop the elevators of a building, even though the silver
surfaces exhibit a certain susceptibility to tarnish. In such contacts, the combination of a strong closing force and high electrical power can break down and
break through a tarnish layer. Phone switching contacts are quite different in
that they must be closed only gently and they carry very little power. Even more
demanding are the connections that must be made and broken and remade, in computer
systems. Here, the power may be measured only in millionths of volts and millionths of amperes, and yet it is absolutely essential that a very small mechanical force will reliably start and stop the electricity flow. No other surfaces
will provide this performance as well as gold.

Indeed, an attempt has been made by the Australian telephone system to use contacts which do not contain gold. This experiment failed since the gold-free contacts caused more noise than could be accepted in telephone voice transmission.

While gold itself is resistant to oxidation and tarnish, it does not invariably hold that a gold plated electrical contact surface will remain free of oxidation. This is particularly true if the gold is plated directly onto a copper support, since copper from the support may diffuse through the gold plate. Then upon reaching the contact surface the copper oxidizes and causes disturbing increases in the contact's electrical conductivity.

Prevention of the undesirable copper diffusion is a major goal of numerous investigations. Problems with base metal diffusion can, for example, be minimized if a diffusion barrier is plated between the base metal support and the gold. Nickel has been conventionally used as this diffusion barrier layer. The use of a nickel barrier layer on copper wire before gold plating has already been mentioned. It has now been found that even better resistance to diffusion is provided by an electroplated layer of an alloy containing approximately equal amounts of tungsten and cobalt, according to Fred I. Nobel and Barnet D. Ostrow of Lea-Ronal Inc.

An alternative approach for reducing base metal diffusion is to use plating techniques which result in gold deposits that are less porous. Many investigators have found that pulsed direct current plating forms dense and continuous gold coatings which are characterized by low porosity. These features make the platings very resistant to base metal diffusion. Pulsed direct current plating is one of a variety of ways now being used by some for preventing the contamination of a gold contact surface with oxidizable copper.

A different type of surface contamination, however, can be quite advantageous under certain circumstances. Gold is extremely suitable for sliding electrical contacts. The excellent performance of these contacts has been contributed to by the presence of a favorable type of polymeric film on the contact surfaces. The polymer is formed from organic vapors by the catalytic action of pure gold. The film acts as a solid thin-film lubricant and insures low friction and low wear without significantly affecting the excellent high contact conductivity of the pure gold. Good electrical contact is maintained since the tiny irregularities on the gold contact surfaces project through the extremely thin polymeric film.

According to W. H. Abbott of Battelle's Columbus, Ohio Laboratories (1979), an alloy of gold with 30% silver catalyzes the formation of this same desirable type of polymeric film. It should be noted, however, that a different and much less desirable kind of polymeric film forms on alloys of gold which contain 30% concentrations of other metals such as copper, nickel, platinum, and palladium. This detrimental polymer can cause considerable wear and, more importantly, undesirably high contact resistance.

W. H. Abbott has also found that a remarkable new gold alloy inhibits formation of the undesirable polymer. This novel alloy also provides the excellent tarnish resistance which is characteristic of pure gold. Much of the improved resistance to polymer formation is attributed to the indium in this 40% gold - 50% silver - 5% cadmium - 5% indium alloy. The contact conductivity for a sliding contact made of this alloy remains high and remarkably stable for over 100,000 electrical contact operations.

DURABILITY

Alloying is commonly used for increasing the hardness and the wear resistance of gold surfaces. Some gold alloys containing less than 5% of base metal can provide excellent wear resistance, while at the same time exhibiting a tarnish resistance which approaches that of pure gold.

Gold resistant to severe wear is utilized on the contact surfaces of make-break contacts. It also provides durability for the contact surfaces of plugs or pins and their sockets. Each year, for example, millions of gold plated plug or pin type connectors are produced by Western Electric for installation in the electronic switching systems which reliably and swiftly transfer phone calls around the world.

Wear resistant gold plates are also utilized on tabs, pads, or plug-like extensions of printed circuit board connectors. Printed circuit boards are often used in movable and replaceable modules. A given circuit board may thus be plugged and unplugged several hundreds of times or thousands of times as the circuitry on that particular board is needed for various specific situations.

Wear resistance of plated golds is important in part so that the contact resistance will not increase due to the exposure of the substrate. As has been shown in the earlier discussion of diffusion, it is important that the gold plate not be worn down to less than a minimum thickness. A too thin gold deposit will permit the diffusion of substrate base metal to the gold surface and thereby permit the formation of undesirable base metal oxides on the surface of the contact.

Electroplated hard golds in which the hardening agent is either nickel or cobalt are very popular. These gold alloys are usually plated from acid cyanide baths although cyanide free solutions have also been developed. The nickel or cobalt concentration in the resulting plated gold alloy is generally less than 1% with values of 0.15% to 0.25% being rather typical.

Lea-Ronal Ltd. with which The Gold Institute/1'Institut de 1'Or is in contact, has developed a new gold alloy which contains cadmium rather than cobalt or nickel. R. T. Hill and K. J. Whitlaw (1978) of this company have found that the excellent wear characteristics of this alloy are comparable with those of an acid bath gold alloy which contains 0.15% cobalt. On the other hand, the stability of the new alloy's contact conductivity is remarkably superior to that of the gold hardened with cobalt. The contact conductivity of the gold-cadmium alloy remains constant at 2.0 reciprocal milliohms for at least 1,000 hours at a temperature of 125 degrees C (257 degrees F). On the other hand, a freshly deposited gold-cobalt electroplate has a comparable conductivity of 2.1 reciprocal milliohms, but after 1,000 hours of operation at 125 degrees C, the conductivity of this conventional electroplate has dropped by 75%.

Iron, rather than cadmium or the conventional cobalt and nickel, is used as the hardening agent in a gold plating solution designated as AURUNA 7000. This gold plating solution is produced by Degussa, a frequent correspondent with the staff of The Gold Institute/1'Institut de 1'Or. Peter Kunz and Hans J. Lübke (1979) of Degussa report that the new acid hard gold can be deposited quite rapidly. With a current density of 6 amperes per square decimeter (0.65 ampere per square foot) this gold is plated at a rate of 50 microinches (1-1/4 microns) per minute. This is 20% faster than the gold plating rate using a similar electrolyte containing cobalt. Wear and hardness tests reveal that the iron hardened gold plate performs comparably to both nickel and cobalt hardened golds.

For certain heavy duty applications it is desirable to use gold contact materials which are more rugged than gold electroplate. A popular type of material is the strip insert bi-metal. A strip of a solid gold alloy is inserted on one or both sides of a bronze, brass, or nickel silver carrier. A typically used gold alloy for this application contains gold, silver and nickel; or gold, silver and copper. The strip insert provides the electrical reliability and durability of solid tarnish resistant gold while mechanical support is provided by selected high modulus metals. Gold insert materials are used effectively and economically in plugs, heavy duty button switches, and sliding springs for television channel selectors. These gold insert materials are produced by DODUCO, Art Wire/DODUCO, and many other fabricators of quality gold products with which The Gold Institute/l'Institut de l'Or is in consultation.

EFFICIENT PRODUCTION

Large numbers of gold surfaced contacts used in communication systems, however, are of the electroplated variety. Modern gold plating systems are called upon to produce such electrical contacts and lead frames in great and increasing volume. Compared to conventional batch processes, the newer automatic systems can often be operated much more rapidly and with lower labor costs. In some cases the improved technology of the newer systems permits twice as many devices as before, to be made per troy ounce of gold, per hour--keeping up with the rapidly growing demand for them.

A continuous and automatic electrochemical system at Western Electric's Dallas Works, for example, now gold plates connector terminals at a high rate, as reported by D. E. Koontz and G. F. Helgeson (1978) and by D. R. Turner and his colleagues (1978) of that company. This improved system selectively plates both soft and hard gold on "947" connector terminals, each of which is half of a 40 or 80 pin connector. The assembled connectors are used to mount and electrically connect circuit packs within Electronic Switch System equipment. The new gold plating system produces 760,000 terminals per day, which is 73% more than the 439,000 terminals per day produced by the previously used batch process.

Also these investigators found that the edges of the plating areas are controlled to within 0.03 inch (about 0.76 millimeter) of the desired boundaries. The variations in the thickness of each gold deposit are only a quarter as large as the thickness variations which result in batch processing.

American Chemical & Refining Company in Waterbury, Connecticut has developed an economical high speed system for applying gold coatings of precise thickness in the form of striping on strip at the remarkable rate of 15 feet per minute.

COMPUTER INGREDIENTS

It should be observed that by no means are all gold deposits prepared by gold plating. For example, thick films formed with gold pastes, and thin films formed with gold inks each have their own special applications.

Each hybrid microcircuit, whether the current LSI (Large Scale Integrated circuit) with 10,000 transistors per chip, or the new VLSI (Very Large Scale Integrated circuit) with more than 50,000 transistors or other elements per chip, contains myriads of conductive paths per square centimeter. For reliability, many of these conductive paths are made with gold. Such reliability is particularly essential for computers, whether they be the hand held variety, which may contain 200 gold surfaces, or the giant industrial variety with gold in thousands of its parts. The cost of the gold is minor compared with the havoc that could result if a computer circuit made with unreliable materials should misplace a decimal point or produce a wrong number in a critical business calculation. Because of this, reliability is ensured by the use of gold.

Engelhard and Johnson Matthey are among the several companies which produce fine selections of compositions for making thin and thick gold films.

Gold is also used in computers to achieve a considerable acceleration in the rate of calculation. Gold which has diffused into a semiconductor diode greatly reduces the minority carrier lifetime. This effect causes the diode to recover very rapidly. Such gold-doped semiconductors can thus be used for extremely rapid switching devices. One recent modification in the gold doping technique gives fast recovering diodes which also have soft, rather than hard turn-off characteristics. Vladimir Rodov of International Rectifier Corporation (1979) has found that the improved gold-containing diode can be turned off without producing adverse high peak voltage oscillations in its circuit.

CONCLUSION

The various technological developments reported here are but a very few of those studied by The Gold Institute/l'Institut de l'Or and continually brought to the attention of industries.

Assistance in preparing this presentation has been provided in part by a number of companies specializing in high quality gold products and systems, which have already been mentioned. Others include the Sel-Rex Division of Oxy Metal Industries Corporation, Technic, Inc., Smith Precious Metals, and Métaux Précieux S.A. Each month the staff of The Gold Institute/l'Institut de l'Or monitors the international technological and patent literature from 37 countries to find new and improved uses for gold in many fields, including jewelry, medicine, catalysis, solar energy, brazing, and architecture. Technical reports and patents found in this continuing search have been included in this presentation of some new developments in which gold is being increasingly used to provide more and better electronics and communications for the world.

REFERENCES

Abbott, W. H. (1979). Frictional Polymer Formation on Precious Metal Alloys. Electrical Contacts -- 1979: Proc. of 25th Annual Holm Conf. on Electrical Contacts. 11-16.

American Chemical & Refining Co. Inc. (Descriptions of gold products), 36 Sheffield Street, Waterbury, Connecticut 06714.

Artwire/DODUCO Corporation (Descriptions of gold products), 9 Wing Drive, Cedar Knolls, New Jersey 07927.

Battelle Laboratories (Descriptions of gold for sliding contacts), 505 King Avenue, Columbus, Ohio 43201.

Degussa (Descriptions of gold products), Postfach 2644, D-6000 Frankfurt 1, West Germany.

DODUCO KG (Descriptions of gold products), Postfach 480, D-7530 Pforzheim, West Germany

Engelhard Industries (Descriptions of gold products), Metro Park Plaza, Iselin, New Jersey 08830 and P. O. Box 340, Aurora, Ontario, Canada L4G 3N1.

Hill, R. T., and K. J. Whitlaw (1978). A New Concept for Electroplated Electronic Contact Golds. Electrical Contacts -- 1978: Proc. of 9th Internat. Conf. on Electric Contact Phenomena & 24th Annual Holm Conf. on Electrical Contacts, 235-241.

International Rectifier Corporation (Descriptions of gold-doped diodes), 9220 Sunset Boulevard, Los Angeles, California 90069.

Koontz, D. E., and G. F. Helgeson (1978). Continuous Plating of ESS Circuit Pack Receptacle Connectors. Part I. A New Approach to High-Speed Selective

Electroplating. The Western Electric Engineer, 22, no. 2, 26-31.

Kunz, P., and J. Donohue (1979). Iron-Containing Gold Deposits from Acid Gold Plating Baths. Metall (Berlin), 33, 254-257.

Kunz, P., and H. J. Lübke (1979). New Acid Gold Electrolytes for High Deposition Rates. Degussa Corporation. 4 pages.

Lea-Ronal Inc. (Descriptions of gold products), 272 Buffalo Avenue, Freeport, N. Y. 11520.

Métaux Précieux S.A./Precious Metals Ltd. (Descriptions of gold products), Av. du Vignoble 2, CH-2000 Neuchatel 9, Switzerland.

Nobel, F. I., and B. D. Ostrow (June 15, 1976). Electrodeposited Gold Plating. United States Patent 3,963,455. 3 pages.

Rodov, V. (February 20, 1979). Process for Manufacture of Fast Recovery Diodes. United States Patent 4,140,560. 4 pages.

Sel-Rex/Oxy Metal Industries Corporation (Descriptions of gold products), 75 River Road, Nutley, New Jersey 07110.

Smith Precious Metals (Descriptions of gold products), 131 Bellows Street, P. O. Box 860, Warwick, Rhode Island 02888.

Technic, Inc. (Descriptions of gold products), P. O. Box 965, Providence, Rhode Island 02901.

Turner, D. R., W. W. Evarts, R. W. Duston, and G. L. Byars (1978). Continuous Plating of ESS Circuit Pack Receptacle Connectors. Part II. Automatic Strip-Plating of Connector Terminals for Electronic Switching System Circuit Packs. The Western Electric Engineer, 22, no. 2, 32-40.

Western Electric (Descriptions of gold components in communications), 222 Broadway, New York, N. Y. 10038.

New Developments in Uses of Silver

Richard L. Davies

The Silver Institute
1001 Connecticut Ave., N. W., Washington, D. C. 20036

ABSTRACT

Sixty-seven examples are given of new developments made in various countries during the past year and a half which appear likely to affect the use of silver in
industry, in the following 20 categories: alloys, batteries & cells, bio-medical,
brazing & soldering, catalysis, ceramics, contacts, conductors, fibers, lubrication & bearings, mirroring & photo-electrics, photography, plating, powders, ster-
lingware & collectibles, tarnish & corrosion resistance, other uses, physical
properties, chemical properties, and production. Numbered references show the
volumes, issues and page numbers of the quarterly publication New Silver Technology
in which details of each development are summarized, with names and addresses of
the developers.

KEYWORD

Silver.

INTRODUCTION

With the widespread use of silver throughout industry, one might assume that the
silver researchers would have no more worlds to conquer. This is not so. From
many countries, new technical papers and patents on silver and silver products are
being produced at the rate of more than ten a day -- more than 5,000, for example,
in the last year and a half since the beginning of 1979. Just during this period,
the staff of The Silver Institute selected 420, in 20 categories which appear most
likely to affect the use of silver in industry. Summaries of each are presented
in the quarterly issues of New Silver Technology which is available by annual subscription from The Silver Institute.

Here are some examples of these recent developments in the 20 categories. The numbered references at the end of this paper show the New Silver Technology issue and
page number for the details of each example.

ALLOYS

Japan has produced two modifications of sterling silver. The first (1), for casting rings and other jewelry, is 2-1/2 times as hard as sterling, and is made of 99.5% silver with nickel and magnesium forming the remainder. The other (2) contains the conventional 7.5% copper, but small amounts of other metals are included to make the alloy resistant to tarnishing.

Metallurgists in the United Kingdom have found that the inclusion of about a third of one percent of silver greatly improves the performance of tough pitch copper for electrical equipment (3), and also the performance of strontium-containing lead for storage batteries (4).

Australians have added .71% silver to an aluminum casting alloy to provide the strength and hardness otherwise available only in aluminum alloys which are expensively wrought rather than simply cast (5).

BATTERIES & CELLS

Great advances have been made in lithium batteries for pacemakers and many other applications, by forming the positive electrode of silver compounds, including silver chromate, silver arsenate, silver pyrophosphate, silver decamolybdate and silver iodate. These new silver batteries have been designed in France (6), U.S.A. (7), and Italy (8), (9), (10), and (11). The Canadians have developed the use of silver boron sulfide together with lithium to make a battery which operates well at temperatures as cold as forty below zero (12).

Advances continue to be made in the design and production of zinc silver oxide and zinc silver peroxide batteries. A new separator made of cellulose impregnated with titanium and silver has more than doubled the number of cycles of charge and discharge possible with a zinc silver oxide storage battery (13).

Advances have been made in the United States in hydrogen - silver peroxide batteries for use in satellites. The hydrogen-silver batteries are forty percent lighter than the previously used hydrogen nickel batteries, and operate effectively even after 4,500 charges and discharges (14).

A new type of battery developed in Japan serves as a precise memory device. The electrodes are silver selenide and silver phosphate, with a silver perchlorate electrolyte. This device has the surprising characteristic that it remembers exactly how much current has been charged into it, and shows this, in a linear way as a corresponding voltage (15).

BIO-MEDICAL

The use of silver sulfadiazine in the healing of burns continues to be extended and perfected in the United States (16). In West Germany, burns and skin infections can now be treated with silver sulfonamide along with cerium nitrate (17); and in India silver sulfamerazine is proving effective for the healing of burns, working somewhat like silver sulfadiazine (18). United States researchers found that polio virus and bacteria can be eliminated from drinking water by passing it through a filter impregnated with silver. This leaves 800 parts per billion of silver in the water, and this is reduced to 50 parts per billion by passing it through an ion exchange resin (19). American doctors can now include silver sulfate in the poly methyl methacrylate cement used in repairing bones, to prevent infection (20).

In India metallic silver sheets are used in radioactive cobalt therapy to prevent attack on the skin (21). Garments for keeping infants warm during surgical operations, and for keeping metal workers cool in front of furnaces, were developed in the United States, using flexible metallic silver mesh to carry warmth or coolness from tubes containing warm or cool water, evenly to the wearer (22). In Japan, silver is now grafted into the molecules of polymers used for wall coverings, floor coverings, and upholstering in hospitals to prevent the growth of bacteria or fungus (23).

BRAZING & SOLDERING

For equipment which must operate at the temperature of liquid helium, 4.2 degrees above absolute zero, many types of brazes were tested in the Soviet Union to provide strong and reliable bonding of the parts. The optimum composition contained 11.8% silver, along with 52.1% copper and 36.0% zinc (24). For joining automobile parts made of powdered iron, American manufacturers began to use a brazing alloy containing 15% silver along with 80% copper and 5% phosphorous (25). For making perfect joints between aluminum and stainless steel for vacuum equipment, and between aluminum and titanium for aircraft, Japanese manufacturers insert thin foils of pure silver between the metals (26) and (27). Power transistors in the United States are now soldered to their silver plated metal packages with a new brazing alloy containing 25% silver, 65% tin and 10% antimony, to provide good electrical conductivity and high strength even with many sharp changes of temperature (28).

CATALYSIS

Advances continue to be made in silver catalysts for oxidizing ethylene to ethylene oxide. U. S. researchers have been able to deposit metallic silver in the form of extremely fine grains so as to increase the yield of ethylene oxide and to permit operation at a lower temperature (29).

In England and Germany, an extremely high yield of formaldehyde can now be obtained by passing a mixture of methanol and air through a honeycomb structure having metallic silver surfaces (30).

Japanese chemists can now produce isoprene for automobile tires, engine mounts and similar applications by a simple, economical process beginning with methanol and isobutene with a catalyst of silver and silicon dioxide (31).

Ozone entering the cabins of jet liners as they fly through the ozone layer of the atmosphere is converted to harmless oxygen by passing it over metallic silver surfaces and other filters as developed in the United States (32).

CERAMICS

United States lens makers have continued their advances in producing glasses which darken in bright sunlight so as to perform like sun glasses, and rapidly become clear again when the light is dimmer. This phenomenon is caused by the presence of about one-half percent silver in the glass. Now bifocals can be made with the silver glass so that the upper part for distance viewing darkens when the light is bright, but leaves the lower part always clear for reading. The same system is applied to ordinary plano glasses as a convenience to automobile drivers, so that they may always have a clear view of the dashboard, but be protected from glare or bright sunlight on the road ahead (33) and (34).

Refrigerated showcases in Japan can now have their glass covers coated with thin films of titanium oxide and an alloy of 95% silver and 5% copper, so that they are transparent to visible light while reflecting heat rays. Thus, they permit the goods in the cases to be seen, but protect the cold merchandise from the heat of the room (35).

CONTACTS

Improved electrical contacts have been developed in the United States for use in circuit breakers carrying very high currents. The contacts are made of 48% metallic silver, 45% tungsten carbide and 7% metallic cobalt (36).

For electric circuits which must be turned on and off thousands of times by contacts which are opened and closed with very little pressure, Japanese electrical engineers have developed a composition made from 92.8% silver, 3% bismuth, 3% tin, and 1.2% zinc. In 20,000 operations, these contacts are 80% more durable and 18 times as reliable in this application as compared with contacts previously used (37).

CONDUCTORS

Circuit boards in the United States are now made with conductive inks containing spherical silver coated glass beads. The conductive strips are cured by and made permanent by a two minute exposure to ultraviolet light, and withstand exposure in tropical environments (38).

A versatile conductive material now developed in the United States is made of silver coated spheres and adhesive ingredients for fabricating conductive gaskets. It combines excellent electrical conductivity with softness, moldability and excellent adhesion to metals (39).

French devices to display digits and letters can now be made by the reversible plating of silver on a transparent conductive oxide electrode. Application of a slight voltage deposits a silver film forming the letter or number. Reversing the potential redissolves the silver, instantly erasing the image. These devices developed in France will operate more then ten million times, at varying temperatures as low as forty degrees below zero without deterioration, and are particularly suitable for use on automobile dashboards (40).

FIBERS

Artificial lawns made with green plastic fibers become grimy in regions of air pollution, since charged particles from the atmosphere are attracted to and held by a charged surface. Japanese chemists have developed a method of keeping the artificial lawns free of this grime by coating the plastic fibers with metallic silver powder. The conductivity of the silver dissipates any charge (41).

In a similar way, carpet makers in the United States make beautiful carpets which do not generate static electricity nor cause sparks when people walk across them. For this, a yarn is used which has a core of colloidal silver dispersed in polyethylene (42).

LUBRICATION & BEARINGS

Work by American manufacturers of large electric generators and motors has now resulted in the development of silver-graphite brushes. When made with 75% silver the brushes can carry a load of more than one and a half million amperes per square foot while sliding over a slip ring at 42 meters per second or over 90 miles per hour. They slide freely with remarkably little wear (43) and (44).

The Japanese Merchant Marine has found a one millimeter thick layer of silver plated onto carbon steel plates which have to be in contact with stainless steel parts of the ship, prevents the serious fretting, or frictional wear which occurs without the silver (45).

MIRRORING & PHOTO-ELECTRICS

United States makers of mirrors for solar energy installations have improved the efficiency of depositing a sheet of pure metallic silver on glass to make mirrors. In a new method a 99.6% deposition efficiency is achieved at a temperature of 100 degrees Fahrenheit, 20 degrees below the temperature previously used. In this method, the three necessary solutions, containing silver, a reducer, and an alkalizer are sprayed on the glass through a single nozzle (46).

In England a solar energy collector has been designed, made with tubes fully coated with heat reflecting indium oxide and partially coated with light-reflecting silver. Even when the intensity of solar radiation is only 200 watts per square centimeter, this device can easily heat water to boiling, from an ambient temperature of 68 degrees F (20 degrees C) (47).

PHOTOGRAPHY

In Bulgaria, photographs, printed circuits and high resolution printing plates can be obtained through the use of a new system. Silver ions are reduced to metallic silver when a thin evaporated layer of arsenous sulfide is dipped into silver nitrate solution. Upon exposure to light, the arsenous sulfide loses its ability to reduce silver ions. Thus, a photographic image is formed by the reduced silver on the areas that have not been exposed to light, but not on the areas which have been exposed to light. For lithography in printing establishments, the arsenous sulfide layer is put on an anodized aluminum substrate (48).

American equipment has been developed for the rapid conversion of printed pages into microfilm and microfiche, and the rapid retrieval whenever needed, in the form of full size sheets like the originals. This new equipment uses photographic emulsions containing small amounts of silver halide and large amounts of silver soaps such as silver behenate. Upon exposure to an image, the film is developed not by the usual chemical solution processing, but simply by dry heating for a few seconds (49).

PLATING

Electrochemists in Germany and the United States have now provided several types of baths for silver electroplating which do not contain cyanide. These are preferred now by many electroplaters because of the severe restrictions in Germany and the United States on traces of cyanide in the effluents. Brilliant silver electroplates of excellent hardness are made in Germany, using silver chloride or silver sulfate along with sodium thiosulfate, and an organic compound such as quinine hydrochloride; antimony is included for tarnish resistance (50) and (51). United States chemists now use a similar cyanide-free solution for directly electroplating a bright silver deposit onto stainless steel (52).

POWDERS

Metallic silver powders with particles smaller than previously available, measuring only 80 angstroms in diameter, have been developed in Poland. They are produced by reducing silver diammine ions with hydroquinone in an alkaline solution. Pastes containing this very fine silver powder are suitable for making the new generation of microelectronic circuits (53). Also in Poland a superb electrically conductive adhesive composition is made with 60 micron size sharp edged silver granules along with silver flakes 10 to 50 microns in diameter (54).

In France, 5 micron size silver powder is combined with silicon polymers to form a material which is an insulator when at rest, but which becomes electrically conductive when pressure is applied to it. It is useful for keyboards and silent switches. The conductivity of a piece of the material 1 millimeter thick is increased 10 million times by finger pressure (55).

STERLINGWARE & COLLECTIBLES

Manufacturing jewelers in Italy developed a way to alter the crystalline structure of silver jewelry surfaces so that they resemble mother-of-pearl. This is done by casting with extremely slow cooling and solidification, followed by polishing and electrochemical etching (56).

Japanese jewelers developed an electrolyte containing sulfamic acid, thiourea, and tartaric acid for electropolishing intricate sterling brooches and other objects to produce beautiful bright surfaces where mechanical polishing is not feasible (57).

TARNISH & CORROSION RESISTANCE

In Switzerland silver jewelry and watch cases are protected from corrosion and tarnish by the vapor deposition of a 100 angstrom coating of manganese and a 2 micron layer of aluminum oxide (58).

Japanese welding and brazing technicians are now using a bonding alloy containing 40% silver, 30% copper, 28% zinc, and 2% nickel to bond stainless steel equipment and to join stainless steel pipes. The silver alloy resists corrosion better than any of the bonding alloys previously used (59).

OTHER USES

Engineers in the United States continue to make advances in the efficiency of introducing silver iodide into clouds to cause rain or snow or to de-fuse hurricanes. One new method is to form the tiny crystals of silver iodide by igniting a composition containing 74.05% silver iodate and 3% hexachlorobenzene, along with aluminum and magnesium powders (60).

In East and West Germany and Japan, improvements were made in silver systems to prevent the radioactive iodine which nuclear power plants generate from coming out into the atmosphere. The improved methods use the silver in various forms, including silver impregnated alumina, silver zeolite, a mixture of silver nitrate, glass fibers and silica gel, and silver powder (61), (62), and (63).

PHYSICAL PROPERTIES

Physicists in the United States have determined that brazing alloys used for equipment at extremely low temperatures should not contain cadmium along with zinc because such alloys become superconductive and upset the instruments used in measuring these very low temperature operations. They found good results, however, with alloys containing 56% silver, 22% copper, 5% tin and 17% zinc (64).

CHEMICAL PROPERTIES

United States researchers can now use silver nitrate for the economical separation of pure ethylene from the cracking reactor gas of a petroleum refinery. The silver nitrate is continuously recycled and it is kept from being reduced, by the addition of small amounts of hydrogen peroxide (65).

PRODUCTION

British metallurgists have developed a modification of the conventional Moebius cell for the electrolytic refining of silver. A dense electrically nonconducting liquid such as carbon tetrachloride is placed at the bottom of the Moebius cell. The silver crystals which form on the stainless steel cathode fall into the carbon tetrachloride and settle to the floor of the tank. There they form a pile for easy removal and washing with carbon tetrachloride which is then distilled off to leave clean, pure, silver crystals (66).

In Belgium a process has been developed for recovering more than 90% of the silver in residues generated in the leaching of zinc ores and concentrates which contain as little as 1% silver. Trichloroethylene is used to dissolve elemental sulfur, and thiourea and sulfuric acid dissolve the silver from the sulfide minerals. The silver is "cemented" on granulated aluminum and the resulting metal is melted and cast to form anodes from which refined silver is produced by the usual electrorefining process (67).

CONCLUSION

It will be seen from this rapid sweep of reports and patents in the past year and a half that dynamic development is proceeding in all these 20 categories relating to silver and silver products. All the indications are that these advances will continue.

Details on the examples given are available in the issues and on the pages of New Silver Technology as listed here. New Silver Technology is available by annual subscription at the nominal price of US$25 for the four quarterly issues of each year.

Richard L. Davies

REFERENCES

New Silver Technology year, issue, and page number.

(1) 1980, Jan., p. 3
(2) 1979, Jan., p. 5
(3) 1979, July, p. 1
(4) 1979, Apr., p. 3
(5) 1979, Apr., p. 2
(6) 1979, Apr., p. 8
(7) 1979, Apr., p. 9
(8) 1979, Apr., p. 10
(9) 1979, Oct., p. 4
(10) 1979, Oct., p. 5
(11) 1980, Jan., p. 8
(12) 1979, Apr., p. 11
(13) 1979, Jan., p. 6
(14) 1979, Oct., p. 3
(15) 1980, Jan., p. 9
(16) 1980, Jan., p. 12
(17) 1979, Jan., p. 11
(18) 1980, Jan., p. 13
(19) 1979, Oct., p. 11
(20) 1979, July, p. 6
(21) 1980, Jan., p. 11
(22) 1980, Apr., p. 8
(23) 1979, Oct., p. 13
(24) 1980, Apr., p. 12
(25) 1979, Apr., p. 15
(26) 1980, Jan., p. 15
(27) 1979, Jan., p. 21
(28) 1980, Apr., p. 13
(29) 1979, Jan., p. 26
(30) 1979, Oct., p. 22
(31) 1980, Jan., p. 18
(32) 1980, Apr., p. 15
(33) 1979, Oct., p. 25
(34) 1979, Oct., p. 26
(35) 1980, Jan., p. 22
(36) 1979, Apr., p. 27
(37) 1980, Apr., p. 28
(38) 1979, Jan., p. 37
(39) 1979, July, p. 27
(40) 1980, Apr., p. 31

(41) 1979, Jan., p. 42
(42) 1979, Oct., p. 36
(43) 1979, July, p. 31
(44) 1980, Apr., p. 38
(45) 1980, Jan., p. 37
(46) 1979, July, p. 32
(47) 1980, Apr., p. 42
(48) 1980, Apr., p. 44
(49) 1980, Apr., p. 45
(50) 1979, Jan., p. 52
(51) 1979, Jan., p. 53
(52) 1979, Jan., p. 54
(53) 1979, Oct., p. 53
(54) 1980, Apr., p. 51
(55) 1979, Jan., p. 57
(56) 1979, July, p. 52
(57) 1980, Jan., p. 58
(58) 1979, Jan., p. 61
(59) 1980, Apr., p. 56
(60) 1979, Jan., p. 65
(61) 1979, Oct., p. 60
(62) 1979, July, p. 57
(63) 1979, July, p. 58
(64) 1980, Jan., p. 67
(65) 1980, Apr., p. 62
(66) 1980, Jan., p. 70
(67) 1980, Apr., p. 66

New Developments in the Use of Gold in Coins, Medals and Medallions

David U. Groves

The Gold Institute/l'Institut de l'Or
1001 Connecticut Avenue, N. W.
Washington, D. C. 20036

ABSTRACT

After centuries of widespread international use of gold as a universally prized money, experiments with monetary systems having less relation to gold resulted in only six nations issuing parts of their legal tender currency in the form of gold coins by 1970. However, by the end of 1979, the trend had reversed itself and 80 nations minted legal tender gold coinage issues to increase their prestige abroad, earn foreign exchange and dampen domestic inflation. The U.S. Treasury Department decision to issue a series of American Arts Gold Medallions is a major recent development in the use of gold in medals and medallions. The increased variety of gold coins, medals and medallions---appealing to many tastes and to as many economic levels---serves to expand the global market for gold coins and other gold pieces.

KEYWORDS

Gold; gold coins; gold medals; gold medallions; Modern Gold Coinage.

Hundreds of years before the birth of Christ, King Croesus of Lydia, Darius the Great of Persia, Alexander the Great, and many other rulers of the ancient world had discovered a new use for some of their precious stores of gold--the coinage of gold coins.

By 1679, all of the nations of Europe were minting and using gold coins while a newly created mint in Mexico City brought the already traditional use of gold coins as a standard of value to the New World.

This extremely beautiful, nearly indestructible metal provided the world with a universally prized money medium until 1914 when, with the outbreak of World War I, Austria-Hungary, France, Germany, Great Britain and other nations halted gold coinage. With the 1930's came the well known experiments with monetary systems having less relation to gold--and the Golden Age of gold coinage went into dormancy.

By 1970, only six countries issued parts of their legal tender currency in the form of gold coins. So, for most of the world, the most recent increased use of gold is also one of the most ancient of its uses--the issuance of gold coinage.

Only yesterday, The Gold Institute presented to its Board of Directors the first copies of its book MODERN GOLD COINAGE 1979. This book reports that during 1979, 80 nations minted 230 legal tender gold coinage issues. That is a 63% increase over the 49 countries that issued gold coins in 1978 and a striking 1,200% increase over 1970's six nations.

According to MODERN GOLD COINAGE--published annually by The Gold Institute with the cooperation of the central banks, ministries of finance and mints of each country-- the amount of gold used in the minting of denominated gold coins in 1979 was 9,905,274 troy ounces, which represents a 1,000% increase over the 832,685 troy ounces of gold which was used for this purpose only ten years ago.

During the past 18 months, the world has witnessed many historic firsts in the field of gold coinage. One of the most important "firsts" occurred here in Canada during 1979 when, for the first time, Canada issued her first gold bullion coin.

This beautiful coin depicts the Canadian Maple Leaf on one side and a portrait of Queen Elizabeth II on the other. It has a face value of $50 (Canadian) and an inscription states that this popular coin contains one troy ounce of pure gold. One million of these coins were issued in 1979 and two million per year are to be minted in 1980 and 1981.

On the other side of the world, for the first time in its history, the People's Republic of China issued 70,000 sets of four 22 karat legal tender 400 Yuan gold coins. Not satisfied with only one "first", the People's Republic of China is now distributing, through International Coins and Currency Incorporated, its first gold coin celebrating the Olympics. Depicting an Alpine skier and the symbol of the Lake Placid 1980 Winter Olympic Games, this gold coin is also the first Chinese gold coin to commemorate an event held in the United States of America. Twenty thousand of these Olympic 916 fine gold coins, each weighing 8 grams and measuring 23 millimeters in diameter, have been minted. Prior to this issue, the rarest modern Olympic coin was the Finnish 500 Markkaa gold coin issued with a mintage of only 18,500 to celebrate the 1952 Helsinki Olympics. It is now trading at fifty times its original cost.

As if an Olympic gold coin issue of 20,000 were not rare enough, the People's Republic of China has borrowed a tradition of the Paris Mint and has minted "Piefort" Proofs of this 250 Yuan gold Olympic coin. These "Piefort" Proofs have been double-struck, on double-thick coin blanks in order to create perfect minting specimens--and only 500 of these double-thick "coin of record" specimens were minted, making them now the rarest of Olympic coins of any kind.

The Olympic Games also sparked another "first" by the Soviet Union, which issued a series of six legal tender gold commemorative coins to promote the 23rd Olympic Games which were scheduled to be held in Moscow in 1980.

But "firsts" were not limited to the largest nations of the world. When the Gilbert Islands became a fully independent republic within the British Commonwealth on July 12, 1979 and changed its name to Kiribati, this tiny group of islands in the Pacific Ocean commissioned the British Royal Mint to strike its first legal tender gold coinage in 22 karat gold. It has a face value of 150 Dollars and depicts a maneaba, a traditional village meeting house on the reverse facing, and a shield with a seabird in flight at the top and a sunset at sea underneath on the obverse.

And in England, for the first time in over 40 years, the British Royal Mint issued a gold Proof Sovereign. This was the first gold Proof Queen Elizabeth II Sovereign to be made available to the public. The 1979 British Proof Sovereign is struck in

22 karat gold, weighs 7.98 grams and measures 22.12 millimeters in diameter.

Since the first gold Sovereign was issued in 1817 by King George III, all Sovereigns have featured the reigning monarch on the obverse facing. Following tradition, the frosted relief obverse of this coin features the famous Arnold Machin portrait of Queen Elizabeth II while the reverse bears Benedetto Pistrucci's "St. George slaying the dragon" as has been the custom since 1887.

Minted in late September, all the coins of the 1979 Queen Elizabeth gold Proof Sovereign's mintage of 50,000 were sold out in North America and in the United Kingdom by the end of October.

And 1979 is also the first year in recent history that 80 nations have seen fit to issue legal tender gold coinage. At this point, we should ask the question, "Why have governments throughout the world returned to the issuance of legal tender gold coinage?"

In brief, they have learned that the issuance of gold coinages produces a variety of benefits to a government.

First, gold coinage increases the nation's prestige abroad by identifying a treasured gold coin with the country or national figure or national concept which the coin commemorates.

Second, it earns foreign exchange, since many of the buyers are from other countries. The premium over the bullion cost of the coin represents clear earnings by the government.

Third, it dampens domestic inflation by letting the country's own citizens put extra cash into the gold coins which they keep, and do not bring to the central bank for redemption. This absorbs domestic cash which would otherwise be spent to fuel inflation.

Over the years, an ever increasing number of people have learned that great benefits also come to the buyer of a legal tender gold coin. They find that they not only have an enduringly beautiful object, but that its value is supported by three price floors:

> ...its face value;
> ...its numismatic value;
> ...and the value of its gold content.

These are some of the factors which have acted to increase the demand for and mintage of legal tender gold coinage.

Other factors were summed up by Herbert J. Coyne, Executive Vice President of J. Aron & Company, Inc., the New York precious metal trading firm, when he said, "Gold is more widely and confidently held than ever before; it is bought and kept largely as a store of value, as a device to protect and preserve purchasing power, and as a hedge against the change in value of money-denominated assets."

Through the global distribution of our MODERN GOLD COINAGE books, THE GOLD NEWS and our other publications, as well as through our continuing contact with the central banks, ministries of finance and mints of each country, The Gold Institute/ l'Institut de l'Or has been providing technical assistance to governments with respect to the profitable issuance of gold coinage.

But what modern gold coins is the public buying? The answers are found in the annual editions of MODERN GOLD COINAGE, published by The Gold Institute.

We find that many people buy gold bullion coins and many others buy gold numismatic coins, while still others buy both gold bullion and numismatic coins.

In a bullion coin, the significant value is the gold contained. At this time, the leading currently minted gold bullion coins are the Krugerrand of South Africa and the Gold Maple Leaf of Canada that was mentioned earlier.

The Krugerrand contains one troy ounce of pure gold in the form of 90% alloy, while the Maple Leaf is of 99.9 percent grade, also containing one troy ounce of pure gold per coin. They are issued at premiums ranging from 2 to 8 percent above the current bullion price. A secondary market for them is constantly maintained so that they can be bought and sold at prices within a few percent of the gold bullion price for the day.

The Krugerrand, with its portrait of Paul Kruger on one side and its portrait of a running springbok or African gazelle on the other, has enjoyed tremendous world-wide demand. In 1979 alone, 4,700,511 pieces were issued and acquired.

Through the export sale of the Krugerrand and the smaller 2-Rand piece, South Africa earned large amounts of foreign currencies and helped South Africa to end 1979 with a $3 billion balance of payments surplus.

While the popularity of the Krugerrand is evident, Mr. Luis Vigdor, Vice President and Manager of the Numismatic Department of the prestigious Manfra, Tordella & Brookes, Inc. in New York, reports that the Canadian Gold Maple Leaf is also finding a large and increasing demand.

Mr. Vigdor explained that although the Canadian Maple Leaf is a bullion coin, it also has a numismatic appeal because Canada limited its mintage to a million pieces during 1979, and each coin is dated.

And, appealing to almost all income levels are the Mexican bullion coins. First minted in 1921, the Mexican 50 Peso coin--with its beautiful winged Angel of Liberty design--has always been very popular with North American and Latin American buyers. Because of the beauty of this coin, for many years its use as a bullion investment has been equalled by its use in the jewelry industry for the manufacture of belt buckle decorations, medallions and charms. But with the gold bullion content value now running over $600 each, most people hesitate to wear too many of these coins as charms on their bracelets.

For the large numbers of people who find it difficult to buy one ounce of gold at today's prices, the Mexican government has five other attractive gold bullion coins--all of 90 percent pure gold--with bullion values running all the way down to less than $30 for the two Peso coin.

Although gold bullion coins receive the most publicity and do make it easier for the issuer or holder to calculate their value, thousands of collectors throughout the world prefer to purchase coins having significant numismatic interest and value.

In the numismatic world, the interest in and value of a gold coin is determined by many factors, including the beauty and condition of a given coin, the coin's face value, the theme presented, the country of origin and the rarity of the coin. These are all values added to the coin's gold content value.

An illustration of this situation is found in a gold coin, weighing just under an

ounce, but rich in its historic value. The first pure gold coin totally minted in
the United States--the Brasher Doubloon of 1787--was sold for $725,000 in an auc-
tion on November 29, 1979 by Bowers and Ruddy Galleries.

Although it is obvious that none of the modern gold coins can match the dollar
value of the Brasher Doubloon, many have generated widespread international interest and demand. Among today's exciting gold numismatic coins are the series of
coins of Hong Kong which mark the Chinese Lunar Calendar years. This year the
$1,000 coin commemorated the Year of the Goat. As in past years it was sold out
within a few weeks after issuance. Then there are Canadian and Soviet Olympic
coins; conservation coins from many nations showing animals, birds and flowers;
coins celebrating the UNICEF's International Year of the Child campaign--and many
others marking events of international or national political or cultural significance.

And how do numismatic coins compare in monetary value with their bullion coin
counterparts? According to one of the largest U.S. coin dealers, Paramount International Coin Corporation of Englewood, Ohio, modern gold coinage compares very
well.

Using data from Paramount's Modern Coin Exchange, which was recently established
by Paramount to match buyers and sellers of modern issue world coins, Paramount
Marketing Manager, Doug Galusha, provided the following examples: The Canadian
$100 gold coin, issued in 1977 to commemorate the Silver Jubilee of Queen
Elizabeth II, sold for an average price of $350 on May 19, or 37 percent above
its gold bullion value on that date. Also in 1977 the Bahamas issued a huge 12
troy ounce 92 percent pure gold coin which now carries an average price tag of
$9,500, or 55% above the value of its gold content.

Moving back only one year to 1976, the popular Canadian gold Olympic $100 coin now
is offered for sale at an average price of $260 which is double its bullion value
of $128.10. The same year, Hong Kong issued its gold $1,000 coin celebrating the
Chinese Lunar Year of the Dragon. And as of May 19, it carried an average sales
price of $800--more than 3 times its gold bullion value.

For those who are interested in having a unique book containing a complete description of every legal tender gold coin issued in the past year, MODERN GOLD
COINAGE 1979 is available for the nominal price of US$10 from The Gold Institute/
l'Institut de l'Or, 1001 Connecticut Avenue, N. W., Washington, D. C. 20036.

With regard to new developments in the use of gold in medals and medallions, a
recent major development is the decision of the United States Government to issue
a series of American Arts Gold Medallions. Only a few days from now, on June 16,
the United States Bureau of the Mint will begin to receive orders for the first
medallion of this series, honoring and portraying singer Marian Anderson and containing 1/2 troy ounce of pure gold. This issue is limited to 1,000,000 pieces.
Then, in the first week of July, another medallion honoring and bearing a likeness
of artist Grant Wood will be available to the U.S. public. This medallion contains
exactly one troy ounce of gold, and the issue is limited to only 500,000 pieces.

In the rest of the world, the medallic arts continue to produce a wide variety of
gold medals and medallions--in France, Finland, other European nations, Mexico and
other Latin American nations, Australia, New Zealand and Canada. It is interesting to note that in the case of both gold medals and gold medallions, the magic of
gold has helped the issuers to realize maximum sales of their gold pieces when the
medals or medallions commemorate interesting and popular themes or subjects.

The increased variety of gold coins, medals and medallions--appealing to many

tastes and to as many economic levels--serve the expanding global demand for gold
coins and other gold pieces. For its part, The Gold Institute/l'Institut de l'Or
will continue to cooperate and consult with the central banks, ministries of
finance and mints of each country to encourage the production of gold coinages
that the nations can be proud of, and that will prove mutually profitable for the
issuer and the holder.

ILLUSTRATIONS

Here are illustrations, courtesy of Manfra, Tordella & Brookes, Inc., 30 Rocke-
feller Plaza, New York, New York 10020, of the beautiful gold coins presented as
slide projections during this presentation.

Austria's 100 Corona gold coin features
a portrait of Emperor Franz Joseph on
its obverse facing.

The reverse of Austria's 100 Corona
gold coin displays the coat of arms of
Imperial Austria and the date 1915.

A standing figure of Emperor Franz
Joseph, with crown and scepter, adorns
the obverse facing of this 1908 gold
100 Korona coin of Hungary.

One of the most beautiful U.S. coins
is the $20 gold St. Gaudens'
portrayal of Liberty.

Benedetto Pistrucci's "St. George Slaying the Dragon" is seen on the reverse facing of British gold Sovereigns.

The obverse facing of the South African Krugerrand shows a bust of President Paul Kruger.

A Springbok is shown on the reverse of the Krugerrand of South Africa.

The reverse of Canada's 1979 pure gold $50 coin displays the Canadian Maple Leaf.

The obverse of Canada's 1979 gold $50 coin bears the Arnold Machin portrait of Queen Elizabeth II.

A full figure of Winged Victory graces the obverse facing of Mexico's popular gold 50 Peso "Centenario" coin.

Jewellery

Platinum in Jewellery

D. E. Lundy

Johnson Matthey, Inc.
Malvern, PA   19355

ABSTRACT

The use of Platinum in jewelry is described in various market areas of the World. Promotions are underway to enhance the consumers image of Platinum jewelry and encourage its purchase. Hallmarking or quality marking laws for Platinum jewelry are discussed.

KEY WORDS

Platinum; Jewelry; Hallmarking; Standards.

INTRODUCTION

Last year at our meeting in Chicago, speaking on this same general subject of "Platinum in Jewelry", I said we were living in the most exciting time in history. That was certainly a prophetic statement even if one considers just the events in the precious metals marketplace. Who could watch the price performances of Platinum, Gold, Silver and Palladium and not be excited. Some of us were so excited, we forgot the old adage "What goes up must come down"!

To those of us interested in the use of Platinum in jewelry, the last year has been one of action and excitement.

Let's take a look at what has been happening.

In any analysis, one has to start with Japan because their purchases of Platinum jewelry still absolutely overshadow consumption in any other country. However, the Japanese proved last year that their passion for Platinum jewelry can be softened by high prices. Since the Japanese consumer pays for his Platinum jewelry at no less than the spot price for the intrinsic value of the metal content of the piece, the price excursions to the $1,000 per ounce level did reduce demand as measured in ounces. Despite this decline in ounces, I am confident the total value of the metal sold at retail greatly exceeded that of any other year. A preliminary estimate for 1979 sales in Japan is 650,000 ounces - this compares to about 800,000 ounces in both 1977 and 1978. If the market price for Platinum stays within 10% of its present level for the rest of the year, it is likely 1980 consumption will again reach a nominal 800,000 ounces.

There is available to the Japanese consumer, a very broad range of delicate and elegantly designed jewelry pieces. The keys to the wide acceptance of Platinum lie in this range of desirable products and, importantly, to its general access-ability throughout the country - these products are highly visible in shops and department stores throughout Japan.

Turning to the Western Hemisphere, on balance, the news has been very good in that Platinum use in jewelry has held up quite well despite the dramatic price increases we've experienced.

In the UK, the total weight submitted for Hallmarking in 1979 declined but the number of pieces marked increased by 13%. Interestingly, comparing the first quarter of 1980 versus that of 1979, Hallmarked Platinum increased 30%, while Gold declined 47% and Silver 59%. So Platinum consumption in the UK for 1980 seems very buoyant despite a significant sales trend reversal in the other jewelry materials. From this, one might generalize that designers have recognized the higher strength-to-weight ratio of Platinum as compared to carat Gold and have created new designs to take advantage of this characteristic. If this is true, then several other nice things happen - the pieces can be less costly as a result of the weight reduction and the constituency has widened into which the jewelry can be sold. Sales fore-casts for the balance of the year are very encouraging.

As in Japan, the increasing availability to the consumer of well-designed Platinum rings, chain and other pieces certainly is fundamental to a growing awareness and preference for Platinum jewelry. Collections of Platinum jewelry are now available in retail shops throughout the UK and a vigorous advertising program on local TV tells the public of this.

In the United States, the Bureau of Mines reported 1979 consumption of Platinum in jewelry was static compared with 1978. Gold consumption was also static while Silver declined about 22%. Of course, because of higher unit prices, the dollars spent on this Platinum undoubtedly set a new record.

The only bleak area is in West Germany where the sales just have not been consistent with the enthusiasm for Platinum that seems apparent among their manufacturers and retailers.

There is an on-going program to promote the use of Platinum in jewelry in four major market areas - Japan, West Germany, United Kingdom and the United States. Funding for the programs in Japan, West Germany and the UK comes from Rustenburg Platinum Mines. Funding for the United States comes from Johnson Matthey.

The aim of the promotion team in each market area is to create a favorable image for Platinum and build a solid base from which to expand its use in jewelry. Every segment of the marketing chain gets attention - manufacturers, designers, retailers and, perhaps most importantly, consumers, are all influenced by some aspect of the promotion.

At this time, and quite understandably, Japan has the most sophisticated and wide-ranging promotion. An annual design competition has evolved into a major event in their jewelry Trade and has produced many designs that have been commercialized with great success - some pieces have even been marketed outside Japan. Because the cost to advertise in glossy consumer magazines is much less expensive in Japan than in other parts of the World, this medium is used extensively to provide con-sumers with many attractive advertisements extolling Platinum jewelry.

In the UK and West Germany, there is a nice balance struck among magazine and TV advertising, retail promotions, fashion publicity and technical service to manufacturers.

Additionally, in the UK, on London's fashionable New Bond Street is THE PLATINUM SHOP, the only retail jewelry store in the World selling only Platinum jewelry. This unique and successful store has been in operation about two years and is operated by a subsidiary of Impala Platinum Mines. When you are in London, take the time to stop by and see their fine and affordable collection of Platinum jewelry.

In each of these three market areas, the formation of a Platinum Guild has been a very important part of the program. The Platinum Guild members are manufacturers and retailers selected because of their interest in and, importantly, commitment to Platinum jewelry. Guild members are able to participate in Trade and consumer advertising and sales promotion campaigns. Their enthusiastic support for Platinum jewelry has been very helpful.

Incidentally, the promotions in Japan, West Germany and the UK are done on behalf of Rustenburg by the local subsidiary of J. Walter Thompson.

In the United States, the promotion has been somewhat less visible to consumers than has been the case in the other countries. We have concentrated our work primarily with the Trade.

Education has been one of our key areas of concentration. We have joined with Gemological Institute of America to sponsor one-day Platinum Workshops. These sessions - there will be ten this year - held in various locations across the country, are attended by jewelry-trade professionals who wish to enhance their skills by acquiring capability in Platinum-smithing. Participants actually make a simple Platinum pendant during the session.

In another program, we provide Platinum metal to selected degree-granting colleges and universities for use in their jewelry arts and metal-smithing courses. Our jewelry industry, through this program, has available an ever-growing cadre of young people already trained in handling Platinum. This is a unique position for an industry which traditionally has been almost forced to hire relatively inexperienced people.

Believing that Platinum has unique qualities which can be exploited by jewelry designers, each year we sponsor a jewelry design competition. The quality of the designs submitted has improved each year and we are quite hopeful that the 1980 Competition will produce designs that are suitable for broad commercialization. I might say that an Ottawa designer won both a Grand Prize and a Runner-Up Prize in the 1979 competition. We are delighted with the broader interest in Platinum now showing in Canada.

Trade advertising and public relations activities are on-going. These create an awareness for Platinum jewelry with the Trade and with the general public. We anticipate that the Platinum jewelry promotion in the United States will soon become somewhat more visible as the result of new programs now in the early stages of planning.

Now to another subject. The marking of Platinum jewelry should be mentioned since marking laws and alloys vary greatly from country to country.

In the UK, there is a rigidly enforced Hallmarking procedure similar to the centuries-old ones for marking Gold and Silver. This requires the use of an alloy containing at least 95% Platinum by weight. The assaying and marking is controlled by the Worshipful Company of Goldsmiths. You may recall Mr. John Forbes, the Assay Master of this August group, was the recipient of the 1979 IPMI Distinguished Achievement Award.

In the European Free Trade Conference countries, there has evolved a common standard for marking Platinum. Under this convention, a Platinum piece made in Norway, for example, could be marked in Norway and then exported to the UK, for example, where the mark would be accepted as having the same value as a UK mark. Their standard requires 95% minimum of Platinum.

In the European Common Market, there is a system for Hallmarking Platinum but it is not uniform nor is it uniformly controlled. Some countries seem to find the system bothersome. I believe there are discussions within the EEC that could lead to a marking scheme that is uniform and enforced.

In Japan, there is a system which is optional - not obligatory. The Japanese Mint Bureau recognizes only four grades of Platinum. These contain either 1000, 950, 900 or 850 parts per thousand respectively of Platinum. I believe there is a growing tendency on the part of manufacturers to mark their Platinum jewelry pieces.

Platinum sold in Canada is marked in conformance with the Canadian Precious Metal Marking Act. This seems to require 95% of the article to be Platinum or an alloy of Platinum and either Iridium or Ruthenium.

In the United States, the marking rules are spelled out in Voluntary Product Standard PS 69-76 approved by the American National Standards Institute on April 18, 1977. There will soon be presented to the Federal Trade Commission a recommended procedure for marking articles made of Platinum. These recommendations will come from the Jewelers Vigilance Committee and will propose 95% as the minimum content to qualify for the Platinum mark. The other 5% is not defined and will permit metallurgists latitude in designing alloys to satisfy specific requirements of the Trade. The Voluntary Product Standard mentioned a moment ago permits as low as 93.5% under certain conditions but specifies the total PGM content must be at least 98.5%.

In summary, I believe Platinum is finding a place in the jewelry markets of the World and these will become an increasingly important outlet for the material during this decade.

A Gold Jeweller's Nightmare: IMF — SDR — FED — London Fix — Krugerrands — LIFO — FIFO — "EMU" — "SWAPS" — M_1, M_2, M_3, Etc. — FIAT

Paul W. Nordt, Jr.

Presentation, International Precious Metals Institute, Tuesday, June 3, 1980
Toronto, Canada.

How simple life was before March 1968! Buying gold for jewelry-making was a simple matter; all one needed was money -- and not very much of it at that: You got a whole ounce of fine gold from the U.S. Assay Office for only thirty-five dollars. It was always the same, for 35 years. If this sheet of paper were made of gold, the thirty-five dollars would buy about a 5 3/4" square. (Sadly, as this is written, those paper dollars buy only about 1½ inches square, making the dollar of the early thirties worth more than fourteen times today's dollar). Oh, to be sure, we had to be licensed by the U.S. Treasury in order to buy gold; and for most of our customers we had to demand an "End-Use" certificate giving an assurance the gold articles we supplied were truly being used for jewelry. We used to wonder what the government did with all the paper; we feel sure there must have been many millions spent on building a place to store them all. In fact, we guessed that nobody ever took time to look at all the data on those forms until one time we made an honest error in filling one out; and, by golly, somebody in Washington caught us. Bureaucracy is wonderful!!!

Anyway, life was easy. For thirty-five years gold was a kind of money, available at the constant price of $35 plus a small handling fee of a few cents per ounce. I say "a kind of" money because for American citizens it was not money at all; the yellow metal was strictly illegal to possess unless in the form of bona fide jewelry or in numismatic coins. To hold gold just for its value was not only against the law, it was considered highly unpatriotic! With The Great Depression gripping the world, we were asked to entrust our benevolent government with the power of managing our money system in order to encourage the return of prosperity. When WW II started in Europe in September '39, we were still wallowing in economic stagnation. No meaningful stimulation came about until we became the "arsenal of democracy" and then finally, with Pearl Harbor, we entered nearly four years of an out-and-out war economy. Although August '45 closed hostilities, in a monetary sense the ravages of war went on for many more years under the Marshall Plan and subsequent aid plans, some of which still go on today. In a very real sense the U.S.A. underwrote its former enemies, giving them the most modern industrial installations while also providing at great cost the military protection that allowed them an environment of low taxes. Thus, with their much greater productivity via modern tools and lower costs through low taxes, much of the world can not out-compete Uncle Sam.

We've taken many decades to awaken to our plight. Over thirty years ago we used our seeming financial power to establish at Bretton Woods the International Monetary Fund (IMF) and the World Bank and the Bank for International Settlements (BIS),

using the power of The Dollar to bolster all world currencies. We pledged to exchange American Dollars for gold, one ounce of gold for thirty-five bucks, to anybody in the world . . . . well not quite anybody. American citizens for some reason or other were not allowed to demand Treasury gold; but anybody else could!

As the years went by, the IMF held regular meetings all around the world, too numerous to mention, lavish affairs providing nice vacations for the participants but
little in the way of solid accomplishments . . . . unless you consider such decisions as the announcement that gold was no longer money and that, instead, the
world's currencies would be backed by something they invented, called Special Drawing Rights (SDR's). These fictitious units were soon dubbed "Paper Gold." These
erudite experts on monetary matters made repeated announcements, always with
straight faces, that gold was being phased out as a monetary material. We had now
reached the stage of intelligent sophistication, in agreement with the patron saint
of fiat money, Lord John Maynard Keynes, that gold was a 'barbarous relic' of ancient money systems and had no part in this brave new world of "fine tuning," managed money supply, and interest rates dictated by The Federal Reserve Board (The
FED). The ultimate cutoff from gold was in August of 1971 under President Nixon.
If ever there were just grounds for impeachment, grounds that would make Watergate
seem as picayune as stealing paper clips, it was the separation, wholly, of our
Dollar from any definition at all. We Americans are so thoroughly brainwashed into
believing that a paper dollar has some valid form of reality and that the so-called
"experts" who manage (finagle is a better word) our money are capable of stabiliz-
ing our currency's value that we, by and large, have let this flagrant violation of
our Constitution go by with hardly a whimper or cry. Since that fatal day in August '71, when The Dollar lost all form of reality, it has been said to be "floating." That term seemed so harmless. More accurately, we might better have said
that is when it was allowed to sink; for in truth, that is exactly what it has done
ever since!

So, that is the background in brief for the Jeweler's Nightmare. More accurately,
it is the background for the nightmare all of us are facing today. A system of
freedom of enterprise can operate healthily only when the government, above all
else, does two things: (The Constitution says it very clearly in the Preamble).
"To provide for the common defense and promote the general welfare." Is it not
disturbingly ominous that our defenses are weaker and weaker (witness the hostages
in Iran) and our general welfare is surely eroding as the common denominator, the
common measure of values The Dollar, is losing ground daily? If one wants to be
more specific concerning the flagrant violation of The Constitution, we can point
to Art. I, Sec. 8, Par. 5 regarding Congress' responsibility over money. That
document admonishes Congress to coin money. It is eminently clear The Framers did
not mean "print," a fact easily made clear with even a perfunctory study of their
times. To make clear that this is violation, we can take note of the remainder of
that clause in Par. 5: (re money) "and to regulate the value thereof." Surely it
takes no circuitous line of reasoning to see that if one is to regulate a value,
that which is to be regulated must, itself, have a definition. How can one possibly regulate something that has no definition, the present plight of our dollars?
Small wonder our dollars buy anything at all!

We jewelers suffered through the "sawtooth" rise from '71 to the end of '74, and
we soon had to learn the nimble art of buying and selling gold that was used in a
product where our "add-on" for fabricating the gold was as little as 10% of the
gold value. If we didn't do it right, we could lose our shirts. The price
changes each day often were far more than the value we added to the metal in the
process of making our products. That seemed great when the price was rising but
horrific when it was low the day we priced an order. Then, to make matters more
scarey, after reaching nearly $200 the end of '74 (a price thought to be fantastic-
ally high at the time), it started a fairly steady descent until the middle of '76.

Our crystal balls did not work too well as we decided to maintain our gold inventory through most of this drop, always guarding against the time when the reversal would signal a new ascent in price. Finally, as the price descended toward $100, we decided to sell off some gold, believing the forecasts that it might go all the way down to about $80. We'd called the turn, in a sense, except that we guessed wrong. Instead of buying back to restore our inventory at a price lower than we'd sold gold, we were in the position of buying back at much higher prices. We'd acted exactly incorrectly as the fantastic rise started, carrying the price to well over $800, a boost of $700 in a period of three and one-half years!

The most recent major trend, commencing in Jan. of this year, had brought the price down again to around the $500 level. It held there within a small percentage variation for over two months, with evidence as this is written that it may again be starting a new burst upward as The Fed now tries as desparately to push down interest rates, just as a few weeks ago it was pursuing its frenetic policy in the opposite direction.

Who can guess what is down the road for us jewelers as our government continues its "Alice in Wonderland" attitudes, attempting to avoid economic realities. With our dollars surely as unreal as "The Emperor's New Clothes" in Hans Christian Anderson's fairy tale, the principal element in our products continues to react, pricewise, to our government's frantic machinations. Paradoxically, many people in and out of government state that gold is "too volatile" a material to be trusted as a standard of value for money. The truth is that gold is simply attempting to respond to the "hare-brained" policies of the U.S. Government as it continues its attempts to monetize paper. Gold's price simply reflects the gasps of a dying dollar.

The prejudice against gold, though unreasoning, is deep-seated in so many minds that its return to favor as a standard for our money is not likely to happen for a long time. The more basic lesson it is crucial we learn mighty soon is the need for some intrinsic standard, possibly other than gold, in order to get our dollars out of the realm of fiction. A columnist in the Wall St. Journal over two years ago suggested a "Cinder Block" standard rather than have no standard at all. Such a facetious proposal would, at least, prevent the uncontrolled running of the printing presses. More facetious, though, that golds ancient role as money will cease.

A whole new vocabulary and new sets of initials have been impressed on our consciousness during the past decade or more. It's no longer sufficient to know that "K" stands for karat, that Dwt. means a twentieth of a Troy Ounce, that "Gr." is a twenty-fourth of a pennyweight, and how gold can be made other colors such as green, red, and white. We now must keep up on such terms as LIFO and FIFO, the meaning of such symbols as M_1, M_2, M_3 etc., that "fiat" is not just an Italian automobile, that an "EMU" is not an African animal, that a London Fix has nothing to do with the drug scene but that an "SDR" truly does seem like the figment of a drug-poisoned mind.

Probably the greatest adversary we jewelers have today is our own government, and our greatest asset in the long run has to be the gold in our inventories, not cash in the bank. There has been an ad running in our trade publications of late telling jewelers their greatest asset is money. Oddly, they show a picture of an old-fashioned silver dollar, not a paper dollar. Despite the recent silver futures fiasco, the old-style "Cartwheel" is worth many times The Dollar it used to define; but if we jewelers try to protect our capital by keeping present-day dollars in the bank, we shall soon learn they will buy less and less of the goods we need in order to do business.

Keeping a business healthy is not an easy task in these times; but aside from a few slips, we try always to maintain our gold stocks. A business with enough capital in the middle sixties to own 1,000 ounces of gold at $35, if it always replaced its inventory on the same market as it sold gold merchandise, can own 1,000 ounces at $500 today. This is a value appreciation, in dollars of $465,000. The jeweler who may be in trouble today is the one who bought his merchandise when gold was, say, $400, applied his normal markup regardless of a later rise in the world market, and after selling it put the money in the bank, waiting for gold to go lower. He is the gambler and often the fool, for the "prudent man" rules applying back when The Dollar had reality are out the window today.

With all the evidence pointing to a much deeper recession than most forecasts, jewelers may be in for rough times. We can look for a disquieting rise in business failures in our industry, many of them the result of an inability to discern economic truths through the smoke screen poured out by the government economists who refuse to recognize the old adage given new life recently by Milton Friedman, "there is no such thing as a free lunch."

I urge jewelers to develop a healthy distrust of all schemes for monetary manipulations and get in the habit of doing one's own thinking. Entrusting such matters blindly to "experts" has led many of us down the garden path toward possible bankruptcy as we continue to think of that piece of paper with the word D-O-L-L-A-R printed thereon as having some form of reality in value. Read the rest of the printing on that "funny" money and see if you can find any pledge or promise by our government telling of its value. It is frightening to discover our dollars are obligating the U.S.A. in no manner whatsoever . . . . . lots of fancy meaningless printing, hardly doing honor to the God in whom we claim to trust.

Problems in Gold Ring Casting

Henry Peterson and Lawrence Diamond

Feature Ring Co., Inc.
New York, New York

ABSTRACT

This article gives a background of the history, development and present state of
the art of gold ring casting. It then deals in more detail with the problems
encountered in model making, mold making, wax injection, treeing, investing, burn-
out and casting. Some causes and solutions are indicated briefly.

KEYWORDS

Gold ring casting; history; development of techniques; problems encountered;
solutions.

INTRODUCTION

This paper will deal with problems encountered in modern mass production methods
of gold ring manufacture. As an introduction, it is of interest to touch upon the
history of gold casting.

Jewelry casting, using gold, has its origins in antiquity, with evidence of its
existence widespread throughout the world. Siberian nomads in Kazakhstan, and
Sumerians at Ur in the Middle East produced beautiful cast works almost 2,700
years ago. The lost wax process was used by the Greeks and Chinese over 2,000
years ago. More recently, starting about 400 A.D., lost wax jewelry casting
reached a remarkable level of sophistication in the New World by the Aztecs, Mayans
and especially by the Indians of the northwest section of South America.

The techniques of mass production and advanced gold metallurgy were mastered by
the Colombian Indians who produced hollow gold casting such as bells with loose,
round hammers trapped within the bell. They understood porosity and had success
in controlling it. Some of the castings were of very fine detail, making the
modern craftsman wonder at their control of temperatures of the molten metal and
casting molds.

It is thought by some that these techniques found their way to Europe and in particular, to Benvenuto Cellini who made written records of lost wax methods. For

nearly four centuries, the lost wax process fell into disuse, with the exception of its use by Carl Faberge who made beautiful jewelry for the czars of Russia.

In 1907, an American dentist, Dr. W. H. Taggert, found a few Italian goldsmiths using the lost wax process in jewelry. He reasoned that it had application to dental restorations and proceeded to use it to make gold teeth and gold crowns.

In the early 1930's, several jewelers started to make gold rings using this process. Most jewelers still used sand casting and cuttlefish casting during this period.

In 1935, a Toronto dental mechanic with an incomplete knowledge of history thought he had an original idea for making jewelry and took out several patents for using the lost wax process. (Apparently, the patent examiner's knowledge of history was equally incomplete.) After several lawsuits, the patents were disallowed, a key piece of evidence being the 400-year-old notes of Benvenuto Cellini.

By 1937, centrifugal casting was being used widely, along with rubber and metal molds which could make many identical reproductions in wax of an original metal model. The need for precision casting during the second world war gave impetus to putting the lost wax process on a scientific basis. Materials and equipment evolved rapidly, with induction melting equipment for centrifugal casting being developed in the 1950's and reached a large use in the late 1960's and early 1970's.

Vacuum assisted casting saw service in the class ring industry in the late 1950's and started to have wide acceptance with the appearance of some improved equipment of European design in the early 1970's.

The latest generation of equipment comes from West Germany which uses pressure assisted vacuum casting concepts.

It is best before we go further to list the various casting methods in use today:

I.   Sand Casting

II.   Cuttlefish Casting

III.   Lost Wax Process and/or Precision Casting

        (1)   Steam and Air Pressure

        (2)   Centrifugal

        (3)   Static Vacuum Assist

        (4)   Pressure-Vacuum Assist

        PROBLEMS

I.   Sand Casting. This is used very infrequently in ring manufacture and is not suitable for high production or where the ring design involves undercuts and small sections.

II.   Cuttlefish Casting. The mold is made from the dried shell of a marine mollusk. One side is hard and the other is soft. The soft side is sanded flat. Two pieces are used. A model to be reproduced is squeezed into each piece from both directions and a sprue is cut into the bottom of the ring shank. The vent lines are scratched in to allow air to escape as the molten metal enters the mold. The model is removed, the two halves are bound together and a pouring gate is opened up leading to the sprue. Molten gold can then be poured in.

The obvious disadvantage of this method is that it can only be used for simple concepts with a minimum of detail and cannot have sharp curves and undercuts. It produces a very rough surface and involves a tremendous amount of hand work, which is not suitable for high production at today's prices for labor. It would seriously retard sales, especially at current gold and diamond prices.

In modern terms, gold ring casting generally follows this procedure:

(a) A wax model is created which is then cast into a metal master, or the model can be made directly in metal. A carefully made sprue is attached. A model is made of material such as brass or silver.

(b) An exact image is made of the metal model in a fusible metal mold or in a rubber mold.

(c) Hot molten or plastic wax is injected into the molds under pressure and then allowed to cool.

(d) The solidified wax reproduction is extracted from the mold. Numerous exact reproductions are made the same way.

(e) The wax reproductions are welded to larger wax supports using the tips of the sprues as the junction point. This is called Treeing. The configuration of the "trees" is a function of the casting process that will follow.

(f) The wax tree assembly is then placed on specially designed rubber mounting fixtures which are removable. An open-ended cylinder is placed over the wax assembly and on to the rubber mounting.

(g) A slurry of investment compound is poured into the cylinder over the wax tree and left to harden. The compound "sets up" and the rubber mount fixture is then removed, exposing the wax sprue tips while snapping off the main wax support. The ghost of the main wax support now becomes the lead gate for the mass of molten gold to enter the mold in the subsequent steps.

(h) The cylinders with the wax patterns are placed in ovens, with the wax part facing down, so that the wax will run out and hot air will rush into the cavities to completely incinerate any remaining wax.

(i) When the investment becomes anhydrous and reaches full tensile strength, the wax removal is complete and the oven temperature is brought to the point that is required for accepting the molten gold alloy.

(j) The molten gold is then introduced into the evacuated hot cavities by various methods - centrifugal force or through vacuum pressure.

(k) The solidified gold is cooled, the investment compound removed and the gold sprues are parted from the gold ring shanks.

Although cast rings are an artistic creation, they can only be made sound and

flawless if their manufacture is considered a precision process. We will now
discuss problems that arise from the original concept to the final cast product.

The ring designer and modelmaker must collaborate to create a product that will
survive the entire ring casting process. The product has to survive assembling,
polishing, stone setting and sizing, and then a pounding when worn by the consumer.

Sprues must be placed on the rings in such a manner so that when gold is forced
or drawn into the hot investment, the flow of metal will reach all points on the
ring with an even, non-turbulent flow. Yet the sprues are to be placed so that
they can be parted, after casting, from the ring. In addition, their location
and size are such that if a ring casting solidifies before the sprue does, an unacceptable freezing depression will end up in the ring.

Most ring casting problems are a function of improper spruing and are the easiest
to solve. It is the balance of the problems that arise that cause bewilderment
and are harder to solve. Assuming that the model is reasonable and the sprues
are made correctly, we now turn our attention to some of the other problems a
caster must face.

FAULTY MOLD MAKING

(a) The rubber is too soft. This will occur if the vulcanization is done at too
low a temperature or for too short a period of time. It may also be caused by
faulty rubber compounding. When pressure is put on the mold as wax is injected
into it, the mold expands, resulting in an overweight ring. The fine detail on
the wax image is lost because the rubber mold, having substandard tensile strength,
allows pieces of itself to be torn off easily as the waxes are extracted.

(b) The rubber is too hard. This will occur if the vulcanization temperature
is a little too high and it is kept in the vulcanizing press a little too long.
It also could be caused by improper compounding. Delicate wax patterns cannot be
extracted as the rubber is too stiff to comply with the force put upon it, and
pieces of the wax are snapped off.

(c) Incomplete wax reproductions. As the wax is injected into the mold, it cannot get into some of the detail because trapped air has nowhere to go. To alle-
viate the problem, vent lines are cut.

(d) Bad mold design. Sections of mold are too weak and distort during injection.
Parting lines in the mold sections are in the wrong places, giving bad seam marks.

FAULTY WAX REPRODUCTIONS

Blow holes and bubbles appear in the wax. This could be produced by moisture in
the pressure lines or air entrained in the wax from excessive wax agitation or wax
pot filling. Foreign matter in the wax comes from poor cleanliness in the wax
melting area or reusing the wax without proper filtration. A lack of parting compound in the mold, which causes wax-rubber adhesion, results in wax reproductions
with cratered surfaces.

Fins and heavy seam marks are a consequence of excessive injection pressures which
will also give an overweight problem. Incomplete wax reproduction is a function
of too low an injection pressure or too low a wax pot temperature. Too high a wax
temperature gives excessive shrinkage as the wax solidifies. This gives distorted
and underweight reproductions.

TREE MAKING

As with all the prior steps, poor technique here will give a host of problems in the casting step. It is essential when fusing the waxes that they should be set in the base of the flask, at the proper angle, and with the correct mass distribution, so that the molten gold will flow into the cavities without retardation. Sharp fragments or sharp corners and stray wax drippings are to be avoided.

All the foregoing will cause investment fragments in the rings, as well as blow holes, porosity, incomplete parts and lumps inside the castings.

INVESTING THE WAX

All sorts of possible problems can arise here and only the strictest controls can avoid them. Failure to adhere to these controls will surely give defects that will show up only in final inspection.

If the tree is not rinsed thoroughly before being invested, stray matter will end up as positive projections on the castings. The wax surface must also be treated to lower the interfacial tension between the investment slurry and the wax to get the full fidelity of the wax reproduction.

Errors or deviations from precision control past this point are not correctible and will introduce excessive repair time, as well as outright rejection in the final product.

The manufacturerof investment compounds makes his material under very stringent conditions and he releases his product to market with individual lot control. He specifies water to solids ratio, age and working time. If his directions are not followed exactly, the caster will pay the piper. The following problems are sure to follow:

(a) Fins and Flash - caused by improper working time, wrong water to solids ratio, and wrong water temperature.

(b) An occasional burst flask - caused by improper working time, wrong water to solids ratio, and wrong water temperature.

(c) Nodules inside and outside on the ring - caused by improper vacuum and/or improper agitation during degassing of the working investment slurry.

(d) Watermarks - caused by too low a water temperature for a correct working time or too low a solids to liquid ratio.

(e) Rough surfaces - caused by improper storage of investment - keeping it too long or exposing it to humid conditions.

After an appropriate time, the rubber fixtures are removed from the set-up investment flasks. Care must be exercised here to make certain that no debris is left behind, for this debris would most certainly be swept into the casting cavities. It is also essential that the setting flasks not be disturbed before the investment is fully set up.

To emphasize the things that can happen in a casting shop, let me relate the following true story:

A supplier of investment was called in to correct a condition where the customer complained that his investment was not glossing off or "setting up". All the control records of the supplier were within specifications, yet the investment did not behave. After some detective work, it was found that the foreman habitually drank grapefruit juice for lunch and was using the can for ladling out water in the investment. The foreman did not know that citric acid in the grapefruit was a powerful setting-up retarder. The problem was solved.

The casting stage is the moment of truth. All the prior preparation is the foundation for the precision filling of the investment mold cavities with molten gold and for the controlled cooling and solidification of the same.

After an hour to an hour and a half, the potted wax patterns are strong enough to be moved to the casting area. The flasks are placed open face down in gas or electric ovens. A programmed heating cycle called Burn-out is initiated which does the following:

1. It eliminates any trace of wax from the investment cavities.

2. Modifies the investment compound which is a blend of a gypsum matrix and a form of silicon dioxide, such as crystoballite, quartz and pure silica, along with other minor additives. Water is driven off in two discreet steps from the gypsum resulting in a sound, hot body, free of cracks and flakes with controlled expansion characteristics. The cycle has a maximum temperature of approximately 1350 to 1400°F. The temperature is reduced to the optimum for the alloy and shape of object and the process used for casting the gold object.

In order for the molten gold to be introduced into the cavities and reach all the fine detail, certain conditions must be overcome:

(a) Natural surface tension of molten gold must be overcome. Molten gold would rather remain a ball than to flow into the cavities. Force must be applied.

(b) A blocking body of trapped air in the path of the flowing gold must be forced out of the way. The investment supplier custom-produces his material with controlled permeability, so that the air can escape through the pores of the investment.

(c) A balance of temperatures of the flask and the molten gold is needed to allow for controlled flow and gold solidification.

There are many means used to force or draw the molten gold into the hot cavities. A simple one is to generate a pocket of trapped steam over a puddle of molten gold and drive it into the flask. Another is to put the hot flask under total vacuum and introduce molten gold under inert gas pressure. This eliminates the trapped air problem in the flask and allows the cleanest part of the gold melt to enter the cavities first. Centrifugal or vacuum assist casting are the ones most commonly used today.

Listed below are a number of things that can go wrong in the Casting Department:

(1) Improper burn-out cycle and excessive burn-out temperatures. If the oven temperature exceeds 1500°F, the gypsum will break down into sulphur dioxide and sulphur trioxide and will result in tarnished, discolored casting. Interruption of the burn-out cycle at certain temperatures can result in hairline cracks and damaged cavities.

(2) Faulty cleanliness, particularly in the open cavity face of the flask. Stray particles of investment can be left in such a position so that the molten gold will sweep these particles into the cavities during the casting process. This results in pock marks and depressions in the final product.

(3) Poor air circulation in ovens will result in carbon residues which will be occluded in the surface area of the casting.

(4) Faulty instrumentation such as worn thermocouples and non-calibrated controls will result in reporting false conditions giving every conceivable type of failure.

(5) Poor air to gas ratios in ovens can lead to a non-oxidizing atmosphere which will inhibit wax elimination and burn-out.

(6) Imprecise gold metallurgy.

   a. Oxides will lead to rough occlusions and improper flow of the molten gold.

   b. Loss of volatile components, i.e. zinc, results in a loss of fluidity of the molten gold, as well as excessive gold percentage in the final product.

   c. Contamination from the shop will lead to stains and other corrosion problems in the final gold alloy.

   d. Substandard alloy components will give porosity, lack of flow and rough casting.

(7) Temperatures of flask and temperatures of the molten gold are very important. These temperatures vary depending on the size and/or weight of the wax reproduction of the model. Heavier waxes require less temperature in the flask and/or less temperature in the molten gold. Lighter waxes require higher temperatures. Items of light weight having small or thin design dimension must receive higher oven temperatures in order to allow the molten gold to fill in the cavities left after wax is eliminated. Excessive metal temperatures will result in the final cooling in the wrong position giving porosity.

(8) Wrong room environment control, i.e. drafts, time delays from oven to casting machines, etc. - any of these will affect the optimum metal and flask temperatures. For example, a cold blast of air in the winter into the open cavities of the hot flask, as it is being inserted into the casting machines, will lower the optimum cavity temperature giving poor results.

(9) Excessive flux can find its way into the cavities along with the gold which will leave indentations on the surface of the casting and too little flux will incorporate metal dross on the surface of the casting.

(10) In the case of centrifugal casting, improper rate of acceleration and improper constant speed before solidification can lead to broken flasks, fins and flash, broken prongs or unfilled fine detail.

(11) If the solidified castings are not allowed to cool a proper length of time and are quenched prematurely in water, massive cracking can result.

Now you can readily understand why gold castings, as we know it today, has been

and still is called the lost wax process or precision casting - precision casting
is the proper name, for one has to be very precise in every step of the lost wax
process to produce a quality product. Ring casting is more than an art - it has
become an exact technology.

REFERENCES

Bovin, M. (1971). Jewelry Making, Bovin Publishing, Forest Hills, N.Y.,72-86.
Erwin, J. E. (1979), Jewelry Casting, Bovin Publishing, Forest Hills, N.Y.
Erwin, J.E. (1979), American Jewelry Manufacturer(27), 82-96.
Gainsbury, P.E. (1979), Gold Bulletin (12),1, 2-8.
Hesch, B., (1979), American Jewelry Manufacturer (27),8, 60-62.
Nielsen, J.P. and Dederer, E.V. (1979), American Jewelry Manufacturer (27),5,
   30-38.
Schultheiss, M. (1979), American Jewelry Manufacturer (27),8, 34-46.
Wald, R. (1975), Introduction to Jewelry Casting, Association Press, New York,
   N. Y., 23-95.
Winkler, N.B. (1967), Jewelers Circular Keystone (CXXXVII), 9, Part II, 80-81
Watchmaker Jeweller and Silversmith (1979(Sept. 51-58.

Static Vacuum Assist in Casting of Silver and Gold Alloys

Albert M. Schaler

James Avery Craftsman, Kerville, Texas

ABSTRACT

Static vacuum assisted investment casting of jewelry and other products is described and compared to centrifugal casting. Four methods are discussed, which give higher productivity, with equal or better quality, and with less skill required.

KEY WORDS

Jewelry; investment casting; static casting; vacuum assist casting.

INTRODUCTION

Investment casting of jewelry parts in precious metals had its beginnings in the 1930's. It was based on dental techniques and it is quite remarkable how few major changes have taken place since its inception. At least 90% of all castings are still made on centrifugal machines; no doubt because this method yields reasonably satisfactory results, and because refinements in the equipment used have diminished the skill necessary for its operation.

There are some inherent disadvantages in centrifugal casting. While many casters are satisfied with using small flasks, especially for gold, others seek to use ever larger flasks in order to achieve better productivity. Unfortunately, considerations of safety and the laws of physics have placed definite limitations on the size of flasks which can be used safely; six inches in diameter by eight inches long would appear to be a reasonable size. The inherent instability of a centrifuge and the comparatively long melting and spinning time needed, especially when using large machines, are other drawbacks.

In seeking to obviate at least some of these disadvantages, some jewelers took another look at an old casting technique - at static casting. Industrial investment casters had long since discarded centrifugal machines and had developed a number of pouring techniques, some of which utilized vacuum assist. Some of these techniques proved to be adaptable to the needs of precious metal casters and at least four distinct types of static casting machines have emerged. Three types will be discussed in some detail while the fourth will be considered only briefly.

This is a system which melts and pours in vacuum. Centrifugal as well as static machines have been built to accomplish this. At best, they produce very fine castings, but they also have some drawbacks. One of them is the very high cost. A centrifuge which melts and casts in vacuum is now quoted at sixty thousand dollars. In order to utilize this equipment fully, it is desirable to use only vacuum melted alloys; these are fairly readily obtainable in Europe but not in the U.S. Most gold alloys sold in the U.S. for casting contain zinc. This renders the alloys impossible for vacuum melting since the zinc will evaporate under vacuum. What can be done is this: a vacuum is obtained before melting begins and then the chamber is backfilled with an inert gas but not to full atmospheric pressure; a costly and rather slow procedure.

Other techniques for static casting are all based on the principle of vacuum assist. A simple definition of the term "vacuum assist casting" might go something like this: Molten metal, which is kept under normal or slightly reduced atmospheric pressure, is introduced into a mold which is kept under a partial or complete vacuum, which will aid in the complete filling of the mold by the molten metal. While either a solid investment mold or a shell mold may be used it is important that the mold material have a fair degree of permeability. The amount of pressure differential between melt and flask is also important, too great a differential may cause damage to the mold, too little will not give perfect filling.

The theoretical advantages of this method are: since there is no centrifugal force involved, one need not worry about flask size. Flasks eighteen inches high have been cast successfully; a firm in Belgium does entire teapots. If enough molten metal is available, there is no limit to the speed with which flasks may be poured; using one technique as many as fifty large flasks have been done in one hour. Metal is poured under very little turbulence; this aspect of centrifugal casting has never been investigated properly, however, it is hard to believe that it helps produce good castings. Other advantages and disadvantages will be discussed together with the various techniques.

The first method to be discussed is quite simple. It requires a casting table and some method of melting metal, usually a gas fired or electric melter which can hold a considerable amount of metal since a number of flasks will be cast in quick succession. The casting table is a steel platform which can accommodate usually four flasks at one time. The flask is placed over a hole which is connected to the vacuum pump. Proper valving is provided for each pouring station. Silicon rubber gaskets assure good contact between the table and the flask.

The actual casting procedure is simplicity itself. The flask is placed over the hole in the table with the flask opening facing up and, either before or after vacuum is applied, the metal is poured into the flask. Normally, four flasks are poured at a time and the vacuum is applied until solidification has set in. The crucible is returned to the furnace for reheating and, as soon as the flasks are solidified, they are removed from the casting table and four more placed in position. There is considerable argument as to when vacuum is to be applied; that is, before or after pouring. Vacuum applied before pouring will almost invariably guarantee good filling of the flask. It does, however, tend to cool the flask rather quickly and the danger of additional oxidation during pouring is considerable. This is, of course, particularly true when casting sterling silver. Applying vacuum after pouring may cause gases to remain entrapped in the flask, resulting in porous if not complete castings. In order to assure a good vacuum being applied to the flask, it is desirable to insert rods along the edge of the flask. These rods, four to eight in number, do not extend the full length of the flask; they are about one to one and a half inches shorter. After inverting, they are withdrawn, leaving hollow places in the flask which provides considerable area for the application of vacuum. This provides better vacuum, especially near that end of the flask where the button is located and where most failure of filling will occur.

The advantages of this method are:
1.   Very high rates of production. As stated before, it is not difficult to pour fifty large flasks an hour with a proper melting setup.

2.   Initial investment costs are very low; a perfectly good setup can be gotten for under $5,000.00.

Some of the disadvantages are:

1. There is not control over the melt or the vacuum, i.e. the pressure differential obtained. Metal is poured through air which is not a desirable practice. The temperature of the metal cannot be controlled to any degree of accuracy, especially when four flasks are poured at one time.

2. Preparing the flask with its vent rods requires more time and it does take away usable space.

3. Pouring speed as well as temperature control depends on the skill of the caster. This method, more than the others to be discussed, requires a good deal of skill. However, if the caster is good, very good results can be obtained.

It is worth noting that this last method works equally well with very large pieces, i.e. statuettes or large belt buckles and with items such as cocktail rings. At present, it is used exclusively for casting copper based non-precious alloys, particularly brass. Very good sterling castings have also been made, however, there is a pronounced tendency towards a heavy firecoat.

The second method is in essence a refinement of the first. Instead of using a standard flask, a flask made of perforated steel is used. On its upper end, a steel collar is welded on. This flask is placed into a vacuum chamber in such a way that the perforated part is inside the chamber and the collar holds the flask in position. A rubber gasket aids in obtaining a good seal between the collar and the vacuum chamber. Again, a vacuum is drawn and the metal is poured into the flask. Because of the perforations of the flask, a much larger area of investment is exposed to vacuum, resulting in a greater and more evenly distributed pressure differential. This method is widely used in casting gold. It is, of course, considerably slower than the first method and it still requires great skill on the part of the caster in order to achieve good results. It should be noted that the molten metal is still poured in a normal atmosphere and, no matter how carefully the melt is protected from picking up oxygen during the melting cycle, it is inevitable that considerable oxygen will combine with the molten material during the actual pour. Actual productivity can be greater than with centrifugal casting depending largely on the melting equipment used. The initial investment is quite modest. It must be noted that the flasks used are very expensive and the welded collars do tend to warp after repeated use. Again, there is little or no control over the actual temperature of the melt nor is there any predictable or repeatable technique of controlling the pressure differential.

The third method represents an attempt to overcome the difficulties inherent in the first two methods. The result is a piece of equipment which is very costly, but produces very good castings and can be operated by a relatively unskilled person. This machine has been used successfully by the author for two years and has fulfilled most of the expectation placed on it.

The machine consists of two major sections. The first to be discussed is the melting section. Melting is done in a graphite crucible placed in an induction coil. The coil is completely insulated, thus effectively protecting the operator. Medium frequency is used for heating. Frequencies from 4 to 20 KC are generated either by a motor driven generator for application requiring 20 KW output or more, or by solid state generators which range from 2 to 10 KW in output. Medium frequency currents import a very good stirring action to the melt, much greater than the high frequencies normally used on the centrifugal machines generally used. As a result, alloying can be carried out very successfully on this machine. The crucible and coil are placed in hermetically sealed housing. There is a quartz window permitting viewing of the melt and this window, which can be swung aside, also serves as the opening through which metal is charged into the crucible. The entire melting assembly can be flooded with inert gas, primarily argon, nitrogen or a nitrogen hydrogen mixture. If desired, the chamber can be evacuated before backfilling with gas. The result is a melt which is largely free of oxides and which does not depend on borax or any other flux for cleanliness. The molten metal flows out through a hole in the bottom of the crucible. This hole is sealed during the melting cycle with a hollow graphite sealing rod which can be raised and lowered from outside the melting chamber. This rod, apart from its sealing function, also contains a thermocouple. This permits accurate measurement of the metal temperature right up to the moment of casting. This is a definite advance over thermocouples used in centrifugal machines which must be removed from the melt before casting. The thermocouple is tied into an electric proportioning controller which will hold a set temperature to within 10 degrees C. plus or minus. Thus, the overheating of metals with all its disastrous consequences is easily avoided and precise. Repetitive melts are routinely achieved.

The second part of the machine, the actual vacuum chamber, is fairly straightforward. It is placed below the melting chamber and it swivels out for receiving the flask. The flask is placed on an air driven platform which will press the flask against the bottom of the melting chamber after the vacuum chamber is swung back in place and locked. A thin silicone rubber gasket is placed over the top of the flask in order to prevent metal splash in the vacuum chamber.

The casting cycle proceeds as follows: Metal is placed in the preheated crucible and heated to a predetermined temperature. When this is reached, the flask is placed on its elevator, the vacuum chamber swung back under the melting chamber and locked, and the flask is driven up against the bottom of the melting chamber and locked.

After the flask is in position, vacuum is applied to the lower chamber and when this vacuum reaches a preset point, normally .4 of complete vacuum, the sealing rod is automatically released. The molten metal is sucked into the flask and will fill since vacuum will be continued until the metal is solidified; normally, about .9 of complete vacuum is achieved.

Within a few seconds after the sealing rod has been raised, it automatically returns to its closed position and, at this point, the still hot crucible may be reloaded. It is important to note that the crucible is never cooled down as is necessarily the case on a centrifugal machine; its controller keeps it at a preset point all during the casting cycle. Melting is very fast; a twenty ounce charge of silver will take little more than a minute to melt so that by the time the previously cast flask is ready for removal from the vacuum chamber, it is time to place the next flask in position. Two persons working together can do up to thirty flasks an hour with this system.

To sum up some of the advantages:

1.   Melting takes place under controlled conditions. The inert atmosphere will prevent rapid oxidizing of the melt and especially in sterling silver firecoat is much reduced. The inert atmosphere also protects the crucible. Upwards of 200 cycles per crucible are not unusual. When casting zinc-rich alloys, it will be found that there is comparitively little zinc loss. The author has cast brass containing over 30% zinc quite successfully.

2.   Fluxing is not necessary.

3.   Automatic temperature control keeps metal from overheating and provides repeatability.

4.   Bottom pouring has always been considered a superior way of pouring since most of the possible contaminants will rise to the top of the melt.

5.   The casting cycle is very rapid and once determined can be repeated at will. The amount of vacuum in the flask at the point of pouring is completely controllable so that thin pieces will be given more vacuum and heavy pieces less.

6.   The castings produced are extremely malleable. We have done very thin sections, as little as .011; wires as thin as .018. Together with the well known Swiss caster, Mr. Hysek, the author has cast complete eyeglass frames. Upon metallurgical examination, they proved completely free of porosity.

7. By using attachments, shot can be made on this machine. Buttons, gates and other scrap can be recycled. The shot produced is quite free of oxides and of water-filled voids. As mentioned previously, alloying can be done quite easily.

The main disadvantages are:

1. The rather hefty capital investment. While there are small machines developed primarily for the dental lab on the market for about $13,000, about twice this sum will be spent on equipment suitable for the manufacturing jeweler. It must be considered, however, that one of these machines can replace at least two centrifugal machines which makes the cost somewhat less painful.

2. Static casting, as such, is considerably less forgiving of mistakes than centrifugal casting. The old adage, "If you can wax it, you can cast it", does not necessarily hold true. "If you can wax it and sprue it properly, you can cast it" would be nearer the truth. Proper gating is extremely important. It is also advisable to use rather heavy tapered center sprues for setting up trees. While such sprues require considerably more metal, they do yield more and better castings. Learning gating and setting up techniques takes time, but it is essential to successful static casting.

3. The machine described here is a sophisticated device utilizing complex electronics as well as compressed air circuits. It also comes from far away and if one lives as the author does, in the heart of Texas, maintenance can become a problem. A good spare parts setup and a lot of ingenuity will be a great help.

4. The last and perhaps greatest disadvantage lies in the high hopes most manufacurers put in such equipment. This machine is by no means an easy road to perfection. It will, no more than any other machine, produce perfect casting from day one. True, it has better controls and greater reliability in repeating a given sequence, but it still must be explained and worked with before it will give its best. It is well known that a good caster can produce good castings on very primitive equipment, and it is equally true that a bad caster will do bad casts on the finest equipment and this rule holds here.

Perhaps a quick comparison between the centrifugal and static casting might be in order here. In most cases, static casting is more productive. The author's firm replaced six centrifugal machines with two static ones with great economies resulting through less personnel.

The cost of investment is arguable. Theoretically, a sophisticated static machine should cost less than an equally sophisticated centrifuge. At the moment, this is not so. Perhaps a little competition will prove helpful.

Both techniques will produce sound casting. Perhaps centrifugal casting is too well entrenched and static casting too new to arrive at a fair conclusion. A few more years should provide better answers.

To close on a personal note . . . . . having cast on centrifugal machines for over thirty years and on static for only two years, would I go back to a centrifugal machine? To quote my late father, "The better is the enemy of the good".

Phase Relations in Cu-Ag-Au Ternary Alloys

H. Yamauchi, H.A. Yoshimatsu[2], A. R. Forouhi and
D. de Fontaine

Department of Materials Science and Mineral Engineering
University of California, Berkeley, CA 94720, U.S.A.

ABSTRACT

Phase relations are studied in the following two portions of the Cu-Ag-Au ternary phase diagram: (a) order-disorder region and (b) eutectic region, two important portions which had been left unclarified throughout the long history of the Cu-Ag-Au system. For Portion (a), we have superimposed theoretically determined "coherent" ternary ordering and miscibility-gap phase boundaries on experimental data of "incoherent" phases. A plausible 300°C isothermal section and one vertical section at 30 at.%Au are given for the phase diagram in which clustering and ordering spinodals are also located. The resulting complex phase diagram can explain quite rationally current transmission electron microscopic observations. For Portion (b), a cooling curve method has been employed to detect the three-phase liquid-solid-solid equilibrium region, and vertical sections are constructed. The position of the eutectic "fold", which extends from the Cu-Ag binary eutectic into the interior of the ternary diagram, is determined in a manner consistent with present observations.

KEYWORDS

Cu-Ag-Au ternary alloys; phase diagram; cluster variation method; order-disorder reaction; spinodal; solid-solid-solid three-phase region; eutectic reaction; eutectic fold; liquid-solid-solid three-phase region; cooling curve method.

INTRODUCTION

Copper-silver-gold ternary alloys have been utilized in various human crafts since prehistoric days. However, only portions of this important ternary phase diagram have been well determined to date. The purpose of the present investigation is to study phase relations in the following two portions of the phase diagram, which have

[1]Research supported by the National Science Foundation (U.S.A.) under Grant Number DMR 79-20405

[2]Now with Mitsubishi Motors Corp., Okazaki, Aichi, Japan

241

been left unclarified or only vaguely determined: (a) order-disorder reaction
region (below 400°C) and (b) eutectic region in the vicinity of the Cu-Ag binary
side (700-800°C).

The importance of order-disorder reactions combined with phase separation for age-
hardening properties in this ternary system has long been recognized (Wise and
Eash, 1933; Yasuda and Ohta, 1979); hence the interest in the Cu-Ag-Au phase diagram. In his phase diagram, Sterner-Rainer (1925) showed a region where a disordered phase coexisted with a Cu-Au ordered phase.   However, this diagram was too
sketchy to be used for quantitative description of phase equilibrium in this
system.   Although Hultgren and Tarnopol (1939) and Raub (1949) studied in detail
the structures of the ordered phases in this ternary system, Uzuka (1977) was the
first to draw boundaries between regions involving ordering reactions in the phase
diagram.   Uzuka's diagram was actually a map of his experimental data and was apparently not intended to do justice to the phase rule.   On the other hand, Kikuchi,
Sanchez, de Fontaine and Yamauchi (1980) calculated a "coherent" phase diagram for
this ternary system using interaction-energy parameters determined to fit only the
binary phase diagrams (Hansen and Anderko, 1958), the Cu-Ag side being modified
according to Cahn's coherent strain energy model (1962).   Although designed to
represent "coherent" equilibria, i.e., those in which all the possible phases differ from each other merely by the arrangement of the constituent atoms, (Cu, Ag or
Au atoms in this case) on a fixed lattice (in this case the fcc lattice of the disordered solid solution), the calculated diagram showed striking similarities with
Uzuka's experimental diagram.   This observation will enable us to superimpose on
the theoretical diagram the experimental data obtained by Uzuka and the "incoherent" miscibility-gap surface reported by Sistare (1973), and consequently to obtain a plausible phase diagram for Portion (a) of the Cu-Ag-Au system.   Theoretical
spinodal surfaces (Kikuchi and colleagues, 1980) will also be adjusted to fit our
plausible phase diagram.   The plausible diagram will then be utilized to explain
observations of decomposition process in Cu-8.2 at.%Ag-32.9 at.%Au (Shashkov,
Syutkina and Rudenko, 1974) and our own transmission electron microscopic observations performed for two alloys; Cu-15.0 at.%Ag-30.0 at.%Au and Cu-35.0 at.%Ag-
30.0 at.%Au.

Many multi-thermal sections of the Cu-Ag-Au liquidus surface have been reported
(Jänecke, 1911; Sterner-Rainer, 1926; Sistare, 1973; Chang, Goldberg and Neumann,
1977), but no detailed observations have been made of the eutectic "fold", a
crease on the otherwise-smooth liquidus surface which must extend from the Cu-Ag
binary eutectic into the interior of the ternary phase diagram.   Interesting topo-
logies are possible for the three-phase liquid-solid-solid coexistence region under
the eutectic fold (Cayron, 1960; Prince, 1966), depending on the shape of the fold.
In the present investigation, phase boundaries involving a liquid phase, that is,
liquidus and solidus surfaces and boundaries of the three phase region, will be
determined by a cooling curve method for various Cu-Ag-Au alloys.   Using these experimental data, possible shapes and location of the eutectic fold will be discussed with the help of thermodynamical considerations.   It should be noted that
there had been only one approximate solidus curve reported on the 14 carat gold
vertical section (Leuser and Wagner, 1953) until Yoshimatsu, Yamauchi and de
Fontaine (1979) determined a portion of the solidus surface in the vicinity of the
eutectic fold in order to find the limit of the single-phase solid solution region
in the area of high gold concentration.

            ORDER-DISORDER REGION

The solid-state portion of the Cu-Ag-Au phase diagram between 240 and 530°C was
calculated by Kikuchi and colleagues (1980).   As mentioned in the previous section,
a single parent lattice, i.e., the fcc lattice, was assumed for all possible

phases. Consequently, ordered phases of CuAu II type with long incommensurate
periods were not taken into account: ordered phases of Cu₃Au type (Ll₂) and CuAu
I type (Ll₀) were considered as well as various disordered fcc phases. The same
interaction-energy parameters were used for the Cu-Au side as those employed in the
Cu-Au binary phase diagram calculation by de Fontaine and Kikuchi (1978). For the
Cu-Ag binary side, the depression of the coherent miscibility gap boundary was
taken into account. In the Ag-Au binary side, an ordering tendency was assumed
corresponding to a critical temperature below 230°C. In the calculation, Kikuchi's
cluster variation method (1951) was employed in the tetrahedron approximation
which was shown to correctly exhibit eutectoid relations between neighboring ordered phases (de Fontaine and Kikuchi, 1978). It should be noted that the cluster
variation method was developed for approximately solving the Ising model, and that
the tetrahedron approximation of the cluster variation method is a higher-order
approximation than the well-known Bragg-Williams and Bethe approximations. A calculated isothermal section at 310°C is shown in Fig. 1, in which various regions

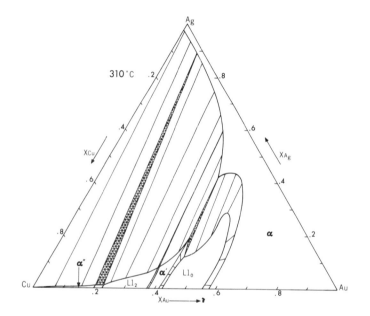

Fig. 1   Isothermal section at 310°C of theoretical coherent
         phase diagram of the Cu-Ag-Au ternary system
         (Kikuchi and colleagues, 1980)

are included: (i) disordered phases α, α' and α", (ii) ordered phases of Cu₃Au
type (Ll₂) and CuAu I type (Ll₀), (iii) coexistence of two disordered phases, i.e.,
(α+α') and (α+α"), (iv) coexistence of a disordered phase and an ordered phase,
i.e., (α+Ll₀), (α+Ll₂), (α'+Ll₀), (α'+Ll₂) and (α"+Ll₂), and (v) coexistence of
three phases, i.e., (α+α'+Ll₀), (α+α'+Ll₂) and (α+α"+Ll₂). The three-phase regions
in an isothermal section must be triangle-shaped, (i.e., tie-triangles) as shown in
Fig. 1. Tie lines in two-phase coexistence regions of (α+α"), (α+Ll₂) and (α + α')
are roughly parallel to the Cu-Ag binary side. Similar features have been confirmed in coexistence regions of two incoherent disordered phases at temperatures
higher than 600°C (Masing and Kloiber, 1940; Ziebold and Ogilvie, 1967; Yoshimatsu
and colleagues, 1979). In the two-phase region of (α+Ll₀), tie lines are observed
to change directions from one edge to the other of the Gibbs triangle.

Experimental data have been reported for the miscibility gap (or solvus surface) by
many investigators (Masing and Kloiber, 1940; McMullin and Norton, 1949; Sistare,
1973; Chang and colleagues, 1977). Uzuka's report (1977) is the unique one that
contains data of equilibrium phases in the order-disorder region of present
interest. All the data mentioned above were obtained by means of powder x-ray diffraction techniques: no experimental information was provided whether the phases
were coherent or incoherent. However, in most cases, investigators annealed specimens for time periods long enough to allow complete evolution to the equilibrium
structures. Therefore, in the present work, we regard these data as representing
"incoherent" phases. Moreover, we consider the linearly extrapolated portion of
the miscibility-gap surface of Sistare (1973) as "incoherent" phase boundaries between two disordered-phase regions.

Simple superimposition of Uzuka's "incoherent" data on the calculated "coherent"
diagram does not show quantitative agreement, nor is it expected to, although similarities are striking. In order to obtain a plausible "incoherent" diagram, we
need to construct phase boundaries so that they may fit not only Uzuka's data
but also the phase relations that the theoretical diagram predicts (Yamauchi,
Forouhi and de Fontaine, 1980). Only relatively minor adjustments are required.
An isothermal section of our plausible "incoherent" diagram thus obtained is shown
in Fig. 2. Not only Uzuka's data, but also a result of Shashkov and colleagues'

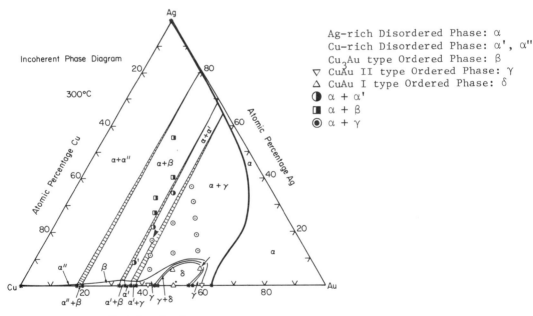

Fig. 2.   Isothermal section at 300°C of our plausible
           incoherent phase diagram of the Cu-Ag-Au ternary
           system. Experimental data were obtained by Uzuka
           (1977), Hansen and Anderko (1958), (*), and
           Shashkov and colleagues (1974), (†).

observation (1974), [marked with a dagger] and phase boundaries on binary sides
[Hansen and Anderko (1958); marked with asterisks] are plotted. As mentioned above,
the phase boundaries between α and α', and between α and α" are linearly extrapo-

lated portions of Sistare's miscibility-gap surface. According to Yasuda and Ohta (1979), the data point marked with an arrow-head represents two-phase coexistence between α and γ (CuAu II) in a 14 carat gold alloy of Cu-15.8 at.%Ag-34.5 at.%Au. The present authors, however, predict that this point should be included in the (α+α'+γ) three-phase region at equilibrium as indicated in Fig. 2. This slight discrepancy can be explained by noting that, experimentally, it is a difficult matter to distinguish three-phase coexistence (α+α'+γ) from two-phase coexistence (α+γ) in this situation, because diffraction spots representing the Cu-rich disordered phase α' are located close to fundamental peaks of the ordered phase γ (CuAu II type). Also, Yasuda and Ohta observed modulated structures in the early stages of phase decomposition of this alloy. This observation tends to indicate that the point with an arrow-head is located inside the clustering spinodal region.

Figure 3 represents a vertical section at 30 at.%Au. In this figure, not only solid-state "coherent" and "incoherent" phase boundaries but also liquidus and

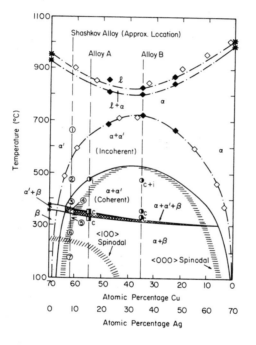

Fig. 3.   Vertical section at 30 at.%Au of our plausible coherent and incoherent phase diagrams including clustering and ordering spinodals. Results of electron microscopic observations are plotted with subscripts c (coherent phases) and i (incoherent phases).

solidus curves are given based on experimental data by Sistare (1973) (◇), Yoshimatsu and colleagues (1979) (◆), and Hansen and Anderko (1958) (*). Theoretical spinodal surfaces for ⟨100⟩ ordering spinodal and ⟨000⟩ clustering spinodal (Kikuchi and colleagues, 1980) are adjusted to fit the plausible phase diagram.

We now compare Shashkov and colleagues' observations (1974) of decomposition process in Cu-8.2 at.%Ag-32.9 at.%Au with those expected from Fig. 3. for a near-by composition of Cu-8.0 at.%Ag-30.0 at.%Au. Shashkov and colleagues reported the following observations: (1) a lower limit of homogeneity of the solid solution was found around 560°C. (2) Between 560°C and 280°C, a Ag-rich phase precipitated by the classical mechanism of incoherent discontinuous precipitation. (3) Below about 280°C, phase separation of the supersaturated solution was observed along with atomic ordering of Cu_3Au type, and fine continuous precipitations were observed in transmission electron micrographs. It is almost straightforward to find the following correspondences between Regions (1)-(3) and Regions ①-⑦ indicated in Fig. 3: (1) ⇔ ①, (2) ⇔ ② + ③ and (3) ⇔ ⑥ + ⑦. The last correspondence implies that the morphology observed in Region (3) may well have resulted from a spinodal mechanism. It must be noted that such narrow regions as Regions ④ and ⑤ can be easily overlooked in experimental observations.

For a further test of the plausible phase diagram construction, we have performed transmission electron microscopic observations on the following two alloys: Alloy A with composition Cu-15.0 at.%Ag-30.0 at.%Au and Alloy B, Cu-35.0 at.%Ag-30.0 at.%Au. Experimental details will be given elsewhere (Yamauchi and colleagues, 1980). Results are plotted in Fig. 3: Subscript c indicates coherent phases while Subscript i represents incoherent phases. Coherency between coexisting phases has been observed at 350 and 320°C in Alloy A and at 470, 350 and 330°C in Alloy B, both being located well inside the "coherent" boundaries. Figure 4(a)

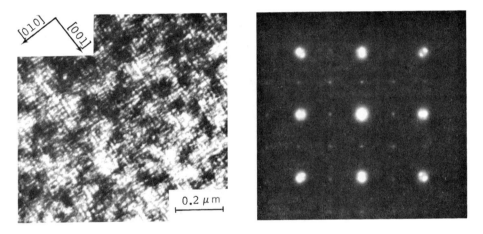

(a)   (b)

Fig. 4.   (a) Transmission electron micrograph of Alloy A,
          Cu-15.0 at.%Ag- 30.0 at.%Au annealed at 350°C for
          5 hours and quenched into iced water, and (b)
          electron diffraction pattern with z = [100]. Super-
          structure peaks due to ordered phase of Cu_3Au type
          are observed.

shows a micrograph taken from Alloy A annealed at 350°C for 5 hours and quenched into iced water. This microstructure exhibits a typical spinodal morphology having a modulation of about 20 nm periodicity along ⟨100⟩ directions. An ordered phase of Cu_3Au type has been confirmed to exist, as shown by the presence of 100 superlattice reflections in a diffraction pattern [Fig. 4(b)] taken from the same specimen. Hence, it may be concluded that the process leading to such coherent microstructures is one of ternary spinodal decomposition into a disordered and an ordered phase.

EUTECTIC REGION

Alloy specimens having compositions in two vertical sections at 10 at.%Au and
20 at.% Au were prepared with 99.9% pure gold and 99.99% pure silver and copper
(Yoshimatsu and colleagues, 1979). Cooling and heating curves were measured for
each specimen with cooling and heating rates of 1-2°C/min and 2-3°C/min,
respectively. Results were recorded with a negative-bias-voltage method
(Yoshimatsu, 1980). Two types of cooling curve were observed above 600°C: (1)
one had a single shoulder and (2) the other exhibited two shoulders, the lower
shoulder being more marked than the upper one. Type (1) curve is a normal one in
which the shoulder gives information about liquidus and solidus temperatures. For
Type (2) curve, Yoshimatsu (1980) attributed the starting point of the upper shoulder to the liquidus temperature, the starting point of the lower shoulder to the
upper boundary of the three-phase liquid-solid-solid coexistence region and the
ending point of the lower shoulder to the lower boundary of the three phase region.
Vertical sections at 10 at.%Au and at 20 at.%Au of the phase diagram have been constructed (Yoshimatsu, 1980). The accuracy of data points was estimated to be within ±3°C. Figure 5 shows the vertical section at 10 at.%Au. A three-phase region

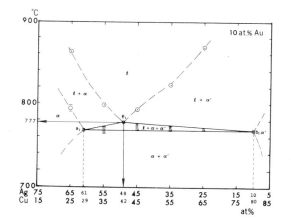

Fig. 5 Experimentally determined vertical section at 10
        at.%Au of the Cu-Ag-Au system. e_1 is the eutectic
        fold point.

($\ell+\alpha+\alpha'$) is shown surrounded by liquid (ℓ), Cu-rich disordered (α') and Ag-rich
disordered (α) phases. The top vertex, e_1, of the region lies on the eutectic
fold. It should be noted that the eutectic fold point e_1 determined by data for
the three-phase region can also be a crossing point of two extrapolated liquidus
curves, as the thermodynamical phase rule requires. This eutectic fold point was
located at 777°C and composition Cu-48 at.%Ag-10 at.%Au. Another eutectic point
was found in the 20 at.%Au vertical section at 790°C and composition Cu-37 at.%Ag-
20 at.%Au. Note that the starting point of the eutectic fold on the Cu-Ag binary
is located at 779°C and composition Cu-60 at.%Ag-0 at.%Au (Hansen and Anderko,
1958). Thus, the eutectic fold has been found to extend almost horizontally from
the Cu-Ag side to the terminal point located beyond the 20 at.%Au vertical plane.
The thermodynamical phase rule requires that, in the vicinity of the binary eutectic on the Cu-Ag side the lowest temperature, T_L, of the three-phase region must be
lower (or higher) than the binary eutectic temperature if the eutectic fold goes
downwards (or upwards) as the Au concentration increases. As is observed in Fig.
5, T_L at 10 at.%Au is at about 770°C, which is significantly lower than the binary

eutectic temperature of 779°C. It is thus most likely that the eutectic fold initially drops down as Au constant increases, contradictory to Lupis' theoretical prediction (1978). Since T_L at 20 at.%Au was raised to about 780°C, a minimum in terms of temperature may be expected for the eutectic fold. Thus, the eutectic fold was found to extend from the Cu-Ag binary eutectic firstly downwards and then upwards to the terminal point located beyond the 20 at.%Au vertical plane.

CONCLUSION

Phase relations of the Cu-Ag-Au ternary system have been studied in (a) the order-disorder region and in (b) the eutectic region of the phase diagram. For Portion (a), a plausible diagram was obtained by superimposing the theoretical "coherent" phase diagram, calculated by Kikuchi and colleagues (1980), on experimental data of "incoherent" phases. A 300°C isothermal section and a 30 at.%Au vertical section was constructed, including ordering and clustering spinodals. Our plausible diagram explained transmission electron microscopic observations of the decomposition process in Cu-8.2 at.%Ag-32.9 at.%Au observed by Shashkov and colleagues (1974). Our electron microscopic observations of Alloy A, Cu-15.0 at.%Ag-30.0 at.%Au and Alloy B, Cu-35.0 at.%Ag-30.0 at.%Au also agreed with the constructed phase diagram. A typical spinodal morphology was observed in Alloy A annealed at 350°C in a two-phase coexistence region between an ordered phase of Cu$_3$Au type and a Ag-rich disordered phase. By means of a cooling curve method, two vertical sections at 10 at.%Au and at 20 at.%Au were constructed in the vicinity of the eutectic fold. Two eutectic fold points were obtained at 777°C and Cu-48 at.%Ag-10 at.%Au, and at 790°C and Cu-37 at.%Ag-20 at.%Au, besides the well-known binary eutectic point at 779°C and Cu-60 at.%Ag-0 at.%Au (Hansen and Anderko, 1958). The observed three-phase liquid-solid-solid region indicated that the eutectic fold extended, from the Cu-Ag binary eutectic, firstly downwards and then rose upwards towards the terminal point located somewhere beyond the 20 at.%Au vertical plane. More experimental work will be required to determine the three-phase region topology with certainty in this ternary system.

REFERENCES

Cahn, J. W. (1962). Acta Metall., 10, 179-183.
Cayron, R. (1960). Etude Théorique des Diagrammes d'Equilibre dans les Systèmes Quaternaires, Institut de Métallurgie, Louvain.
Chang, Y. A., D. Goldberg, and J. P. Neumann (1977). J. Phys. Chem. Ref. Data, 6, 627-629.
de Fontaine, D. and R. Kikuchi (1978). In G. C. Carter (Ed.), Application of Phase Diagrams in Metallurgy and Ceramics, Vol. 2, NBS Special Publication 496. pp. 999-1026.
Hansen, M. and K. Anderko (1958). Constitution of Binary Alloys, McGraw-Hill, New York.
Hultgren, R. and L. Tarnopol (1939). Trans. AIME, 133, 228-238.
Jänecke, E. (1911). Metallurgie, 8, 597-606.
Kikuchi, R. (1951). Phys. Rev., 81, 988-1003.
Kikuchi, R., J. M. Sanchez, D. de Fontaine and H. Yamauchi (1980). Theoretical Calculation of the Cu-Ag-Cu Coherent Phase Diagram. Acta Metall. In press.
Leuser, J. and E. Wagner (1953). Z. Metallkd., 44, 282-286.
Lupis, C. H. P. (1978). Metall. Trans. B, 9B, 231-239.
Masing, G. and K. Kloiber (1940). Z. Metallkd., 32, 125-132.
McMullin, J. G. and J. T. Norton (1949). J. Metals (Metals Trans.), 1, 46-48.
Prince, A. (1966). Alloy Phase Equilibria, Elsevier, New York.
Raub, E. (1949). Z. Metallkd., 40, 46-54.
Shashkov, O. D., V. I. Syutkina and V. K. Rudenko (1974). Phys. Met. Metallogr. (USSR), 37, 94-100.

Sistare, G. H. (1973). In Metals Handbook, Vol. 8, American Society for Metals, Ohio, pp. 377-378.

Sterner-Rainer, L. (1925). Z. Metallkd., 17, 162-165.

Sterner-Rainer, L. (1926). Z. Metallkd., 18, 143-148.

Uzuka, T. (1977). J. Japan Soc. Dent. Apparatus Mater. [Shika-Rigaku Zasshi], 18, 67-72.

Wise, E. M. and J. T. Eash (1933). Trans. AIME, 104, 276-307.

Yamauchi, H., A. R. Forouhi, D. de Fontaine (1980). A Plausible Phase Diagram for the Solid State Portion of the Cu-Ag-Au System, In preparation for publication.

Yasuda, K. and M. Ohta (1979). Age-Hardening in a 14 Carat Dental Gold Alloy. Proceedings of 3rd International Precious Metals Conference, International Precious Metals Institute, New York.

Yoshimatsu, H. A., H. Yamauchi and D. de Fontaine (1979). Experimental Study of Phase Transformations in the Cu-Ag-Cu Ternary Alloy (I), UCB Research Report: UCB-ENG-DDF 7901, University of California, Berkeley.

Yoshimatsu, H. A. (1980). Eutectic Phase Relations in the Gold-Silver-Copper Ternary System, M. S. Thesis, University of California, Los Angeles.

Ziebold, T. O. and R. E. Ogilvie (1967). Trans. AIME, 239, 942-953.

Dental

The Corrosive Attack of Gold-Based Dental Alloys

R. M. German[*], D. C. Wright[**] and R. F. Gallant[**]

[*]Materials Engineering Department, Rensselaer Polytechnic Institute, Troy, New York 12181

[**]Research Division, J. M. Ney Company, Bloomfield, Connecticut 06002

ABSTRACT

Systematic study is given to gold-based dental alloys using potentiodynamic corrosion scans. Characteristic corrosion behavior is demonstrated for both Ag and Cu; thus corrosion susceptibility is characterized by the magnitude of the current density peaks associated with these elements. Examination of several dental alloys shows that nobility is a dominant factor in determining corrosion resistance. However, control of microstructure provides a means of enhancing corrosion resistance for the low nobility alloys. The implication of the corrosion data are discussed with specific reference to design requirements for low nobility dental alloys.

KEYWORDS

Gold alloys; dental alloys; corrosion; potentiodynamic scans; nobility; alloy design; cytotoxicity; microstructure.

INTRODUCTION

The progresses in metallurgical research have lead to a wide selection of Au-Ag-Cu based alloys for dental applications. Several low-gold formulations are available with mechanical properties equal to those found in high-gold content alloys. While the economic attraction of these alloys is obvious, there are some serious drawbacks. For example, the high nobility alloys prove easier to cast and work. More importantly, with lower nobilities there are serious questions concerning long-term biocompatibility. The low-gold compositions are characteristically high in either silver or copper, two elements with evidenced cytotoxicity. The alternative low cost nickel base alloys are of concern because of unanswered questions concerning carcinogenicity. Hence, there is an effort underway to develop several laboratory tests for dental alloy stability in the oral environment (tarnish, corrosion, cytotoxicity, chemical leaching, etc.). This paper reports the development and application of a potentiodynamic test for corrosion resistance. Four specific points are covered in the presentation; the corrosion test methodology, data for several gold based dental alloys, illustration of the dominant role of nobility, and discussion of benefits possible through control of alloy microstructure. We conclude with a discussion on the implications of the data and an outline of corrosion-based design requirements for dental casting alloys.

EXPERIMENTAL APPROACH

Potentiodynamic polarization techniques were used to monitor the corrosion behavior of cast alloy samples in deoxygenated 1% NaCl electrolyte (Wright, German and Gallant, 1980). For these studies, chloride environments appear to be the most meaningful; Hoar and Mears (1966) have shown a strong effect of chloride ions on the corrosion of biometals. The corrosion cell consisted of the sample as an anode, platinum as a cathode and a saturated calomel reference electrode. The current density was measured versus the impressed voltage from -600 to +400mV in both anodic and cathodic scans at 2mV/s. Current density peaks for copper and silver were determined for several alloys as detailed elsewhere (Wright, German and Gallant, 1980; Sarkar, Fuys and Stanford, 1979a, 1979b). The peak current densities were monitored against nobility, alloy type and microstructure using the anodic scan peak at approximately -125mV for copper and the cathodic scan peak at approximately 0mV for silver. Scanning electron microscopy and energy dispersive x-ray analysis were used to link the corrosive peaks to the specific elements and alloy microstructure.

Altogether 28 Ag-Au-Cu experimental alloys and 11 commercial crown and bridge alloys have been examined. The main findings of these experiments will be illustrated with the seven alloys listed in Table 1. Included with the nobilities are the summary data for the corrosion current densities at the Cu and Ag characteristic voltages (anodic and cathodic scans respectively).

TABLE 1   Dental Alloy Corrosion Data

Alloy	Nobility wt.%	Nobility at.%	Composition, at.% Au	Ag	Cu	Pd	Pt	Zn	Current Density, $\mu A/cm^2$ Cu peak	Ag peak	Max.[*]
1	85	73	67	18	8	6	0	0	1	1	1
2	76	56	49	16	24	5	2	4	0.4	4	8
3	65	47	43	33	17	4	0	3	1	1	3
4	52	30	21	8	60	8	1	2	1	2	20
5	46	25	20	8	58	4	1	9	1	1000	1000
6	39	31	8	42	26	22	1	1	2	90	90
7	25	23	0	53	21	23	0	3	3	6000	6000

[*]Maximum current density between +200mV and -200mV.

RESULTS

The test methodology described briefly above gives basis for measuring the relative Cu and Ag corrosion susceptibilities of this alloy class. Within the Ag-Au-Cu based dental alloys, we have succeeded in reproducing potentiodynamic scans with the deoxygenated 1% NaCl. Following a parametric study, it was established that the test method provides a quantitative basis for rank ordering dental alloys for their corrosion resistance. With such a standardized test, it becomes possible to isolate the effects of specific parameters on corrosion behavior.

The data shown in Table 1 give examples of the behavior observed in this study. There is a generalized increase in corrosion resistance with increased nobility. However, there is evidence that more than just nobility is involved in determining the corrosion rate. For simple alloys based on binary or ternary Ag-Au-Cu combinations there is a predominant role for nobility. This point is illustrated in Fig. 1 where the current densities are shown as functions of nobility (i.e., at.% Au). However, the data in Table 1 evidence corrosion variations for dental alloys (Ag-Au-Cu-Pd-Pt-Zn) which are not totally determined by nobility.

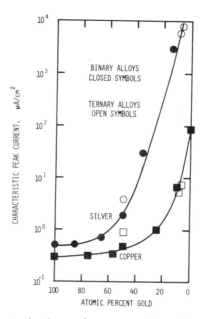

Fig. 1. Characteristic peak current densities (corrosion)
for copper and silver measured by potentiodynamic
scans for alloys of various gold contents.

Examination of alloy microstructure provides important additional information concerning the distribution of alloying elements (Eick and Hegdahl, 1968). For example alloy 5 is similar in chemistry to alloy 4 except that the former is two phase. In alloy 5 there is a segregated silver rich second phase which is susceptible to chloride formation. Examination of the surface following the anodic scan reveals silver chloride has formed at the second phase. Figure 2 shows a scanning electron

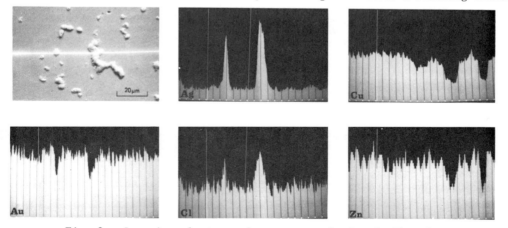

Fig. 2. Scanning electron microscopy analysis of alloy 5
following the forward scan. The energy dispersive
x-ray data for Ag, Cu, Au, Cl and Zn taken from the
indicated line scan identify the corrosion product
as a silver chloride.

micrograph of the surface following the forward scan; energy dispersive line profiles evidence the silver chloride. Similarly, it was found that alloy 7 was two
phase with a typical segregated, dendritic microstructure. Studies by Gallant,
German and Wright (1980) have demonstrated that alloys with a tendency towards
segregation or second phase formation are prone to high corrosion rates. Furthermore, these same alloys can be improved by either homogenization heat treatments
or rapid cooling from the molten state.

DISCUSSION

Potentiodynamic polarization has been applied to determining a relative corrosion
rating for gold based dental alloys. Through measurements of the current densities
at various points on the potentiodynamic scan it is possible to rate corrosion
susceptibility. Such an approach is important to determining alloy acceptability
and as such will aid in future alloy development. With the obvious trends towards
lower nobility, it will be necessary to more fully characterize alloy corrosion
resistance by the techniques outlined in this report.

Examination of Fig. 1 shows that corrosion of simple gold-based binary and ternary
alloys is highly dependent on nobility. The noble metal content on an atomic basis
is a leading indicator of corrosion resistance. Specifically, with noble metal
contents above approximately 55 at.%, there is little evidence of corrosive attack.
Alternatively, below this nobility, alloy microstructure becomes a variable of
concern. Generally, with the same nobility single phase alloys will be superior
in corrosion resistance to two phase alloys. For a two phase alloy some improvement in corrosion behavior is possible through rapid cooling from the melt. Alloys
with fairly low nobilities are inherently more prone to corrosive attack and hence
microstructure control has a more pronounced influence.

In terms of alloy corrosion resistance, this quantitative study provides a basis
for rank ordering alloys. From Table 1, the following rank ordering of alloys is
offered (best to worst) 1,3,2,4,6,5,7. The first three of these have nobilities
over 55 at.% and hence corrosion problems are not anticipated based on this study.
The last two alloys (5 and 7) are dual phase, segregated alloys with less than 55
at.% noble metal. Example optical micrographs for these two alloys are shown in
Fig. 3. Such alloys pose some difficult questions with respect to long-term oral
stability and should be used with caution. Thus based on these corrosion measurements we can formulate the following design guidelines for gold-based dental alloys: 1) nobilities above approximately 55 at.% are most desirable, 2) single
phase alloys are preferred at all nobilities, 3) casting segregation should be
minimized through grain refinement and rapid solidification, and 4) alloys with
less than 55 at.% noble metal and two phase microstructures are undesirable for
dental use. Concurrent studies involving tarnish resistance (German, Guzowski and
Wright, 1980; Wright and German, 1979) further stress the necessity to maintain
single phase microstructures for optimal performance of low nobility dental alloys.

CONCLUSIONS

Alloy nobility on an atomic basis provides the best overall measure of corrosion
resistance for the gold-based dental alloys. Generally single phase microstructures with minimized segregation through grain refinement, rapid solidification or
heat treatment perform the best in potentiodynamic corrosion tests. At low nobilities (below approximately 55 at.%), the differences in alloys become significant.
Small chemistry differences can produce microstructural variations which greatly
increase the corrosion rate. However, the overall benefits possible through control of microstructure are secondary in importance to nobility.

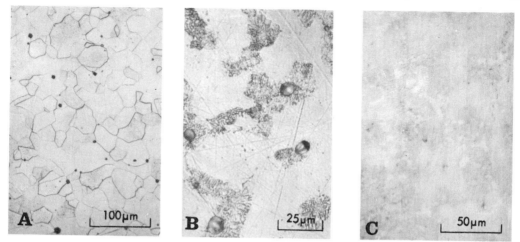

Fig. 3.   Comparison of microstructures for alloys 3, 5, and
7 (A, B, C respectively). Alloy 3 is a high
nobility, low corrosion alloy with the uniform
microstructure deemed appropriate for gold-based
dental alloys. Alloy 5 is a two phase structure
with an eutectic grain boundary morphology. Alloy
7 is segregated, two phase, low nobility alloy with
substantial corrosive attack.

REFERENCES

Eick, J. D., and Hegdahl, T. (1968).   Segregation in dental gold casting alloys.
J. Dental Res., 47, 1118-1127.

Gallant, R. F., German, R. M., and Wright, D. C. (1980). Effect of investment
temperature on microstructure and corrosion behavior. In Proc. Dental Materials
Group, 1980 AADR Annual Meeting, available Dr. J. D. Eick, Dental School, Oral
Roberts Univ., Tulsa, Oklahoma.

German, R. M., Guzowski, M. M., and Wright, D. C. (1980). Color and color stability as alloy design criteria. J. Metals, 32, 20-27.

Hoar, T. P., and Mears, D. C. (1966). Corrosion-resistant alloys in chloride
solutions: materials for surgical implants. Proc. Royal Soc. (Ser. A), 294,
486-510.

Sarkar, N. K., Fuys, R. A., and Stanford, J. W. (1979a). The chloride corrosion
of low gold casting alloys. J. Dental Res., 58, 568-576.

Sarkar, N. K., Fuys, R. A., and Standord, J. W. (1979b). The chloride corrosion
behavior of silver-base casting alloys. J. Dental Res., 58, 1572-1577.

Wright, D. C., and German, R. M. (1979). Quantification of color and tarnish
resistance of dental alloys. In Proc. Dental Materials Group, 1979 AADR/IADR
Annual Meeting, available Dr. J. D. Eick, Dental School, Oral Roberts Univ.,
Tulsa, Oklahoma.

Wright, D. C., German, R. M. and Gallant, R. F. (1980). Copper and silver corrosion behavior in crown and bridge alloys. In Proc. Dental Materials Group,
1980 AADR Annual Meeting, available Dr. J. D. Eick, Dental School, Oral Roberts
Univ., Tulsa, Oklahoma.

Physical and Clinical Characterization of Low-Gold Dental Alloys

K.F. Leinfelder, R.P. Kusy and W.G. Price

Dental Research Center,
University of North Carolina,
Chapel Hill, NC 27514 (USA)

ABSTRACT

Due to the escalating cost of precious metals a number of low-gold alloys have been introduced to the dental profession. In spite of their general acceptance little information is currently available regarding their physical and mechanical properties. This study deals with the effect of composition on certain mechanical properties with particular emphasis on burnishability.

KEY WORDS

Low-gold alloys; mechanical properties; burnishability; corrosion.

INTRODUCTION

The use of gold as a dental restorative material dates back several centuries. Its early popularity was based on the fact that this metal could be readily worked and adapted to tooth structure. Furthermore, unlike most other metals available at that time, gold was resistant to tarnish and corrosion in the oral cavity, i.e., noble. The advent of the precision cast technique near the turn of the twentieth century prompted the development of several gold alloy compositions. Over the years the American Dental Association has written a specification which standardizes not only their precious metal content but also their mechanical properties (American Dental Association Specification No. 5, 1966). While the specification provides standards for four different types of alloys, only two are currently applicable to dental practice. The average compositions of these are referred to as Type II and III alloys, and are shown in Table 1.

Each of the elements used has a specific function. Gold provides tarnish and corrosion resistance as well as ductility, burnishability, and heat hardening characteristics. Most references suggest that alloys containing less than 75 weight percent gold or 18 kt. will tarnish and corrode in the oral environment (Paffenbarger, Sweeney, and Isaacs, 1932; Report of Conference on Dental Inlay Gold Alloys at National Bureau of Standards, 1931; Wise, Crowell, and Eash, 1932). Copper in combination with gold contributes strength to the overall

259

TABLE 1   Typical Compositions of ADA Type II and III Gold Alloys

Element	Type II (inlay - onlay), %	Type III (crown & bridge), %
Au	77.0	72.0
Cu	8.0	10.0
Ag	12.0	13.0
Pd	2.5	3.0
Pt	0.0	1.0
Zn	0.5	1.0

composition. Moreover, when added in sufficient amounts it renders the alloy heat hardenable. Silver plays a number of important roles. First of all it is important in controlling the alloy color. Minor amounts are effective in neutralizing the red color imparted by copper. Furthermore, silver contributes to ductility and prevents gold-copper alloys from excessive ordering. When subjected to a hardening heat treatment, 18 kt. gold-copper alloys undergo grain boundary fracture. Typical examples of this phenomenon are illustrated in Figs. 1 and 2 after aging at 350°C for 30 minutes. The first shows a cast bar of the alloy, while the second illustrates a clinical restoration of the same. Fortunately, the addition of five to six weight percent silver eliminates this unacceptable type of failure. Platinum and palladium contribute to solution hardening of the alloys and tend to whiten the color of the casting. In addition, in small quantities both these elements elevate the fusion or casting temperatures. Low percentages of zinc or indium are frequently added to increase the fluidity of the alloy during casting. Finally, for about the last 15 years most dental gold alloys have contained 50-100 parts per million of a grain refiner, either iridium or ruthenium. Either element reduces the grain volume by as much as 125 times.

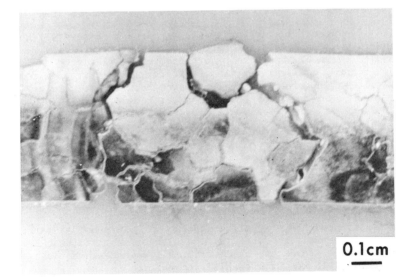

0.1cm

Fig. 1.   Optical photomicrograph of a cast bar of 18 kt. gold-copper alloy. Heat treatment consisted of solution annealing at 700°C for 2 hr followed by aging at 350°C for 30 minutes.

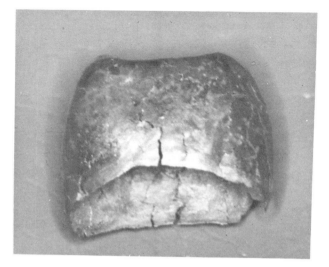

Fig. 2.   Photograph of clinical casting after aging at 350°C for 30
minutes. Note cracks (at grain boundaries) radiating
throughout the casting.

Due to the recent speculation in gold and other precious metals, these
traditional casting alloys have become extremely expensive. While a full cast
crown had an inherent value of only $30.00 less than three years ago, its value
has escalated to $100.00 or more. As a result, a number of manufacturers have
introduced a variety of casting alloys to the dental market that contain reduced
gold contents. The percent gold of most of these new alloys ranges between
40-60% with most of them approximating 50%. To maintain the same level of
corrosion resistance associated with the traditional casting alloys, the
manufacturers have increased the palladium and silver contents. Instead of one
or two percent palladium, the reduced gold alloys contain 3 - 9 percent
palladium. The silver content by comparison has been increased by 200 to 300
percent. None of these newer alloys contain platinum. Due to the elevated
palladium content, platinum additions would excessively elevate the casting
temperature, increase the hardness, and further whiten the alloy.

Although there are as many as thirty or forty of these alloys on the market,
little information is available regarding their physical and mechanical
characteristics. Moreover, there is no information regarding their resistance to
tarnish and corrosion in the oral environment. Consequently, this study
investigates the physical and mechanical characteristics, associated solid state
transformations, and clinical performance of these alloys.

EXPERIMENTAL

Materials

The proprietary alloys, their respective manufacturers, and gold contents are
given in Table 2.

TABLE 2 Proprietary Alloys Evaluated

Alloy	Manufacturer	Gold Content, wt.%
*Modulay	Jelenko	77
Maxigold	Williams	60
Mowrey #120	Mowrey	50
Midigold	Williams	50
Midas	Jelenko	46
Mowrey #46	Mowrey	42
Forticast	Jelenko	42
Minigold	Williams	40

*Control

A conventional American Dental Association Type II gold alloy was included for the purpose of comparison. In addition to the proprietary compositions, eight experimental alloys were premelted in a graphite crucible and homogenized in a vacuum from high purity silver, copper, and palladium (four nines-five) and gold (three nines-five) (Table 3).

TABLE 3 Experimental Alloys Evaluated

Alloy	Gold Content, wt.%
UNC #6	49.9
UNC #1	48.0
UNC #2	48.0
UNC #3	48.0
UNC #4	38.4
UNC #7	36.8
UNC #5	28.8
UNC #8	27.6

The gold-silver-copper ratios of both groups of alloys are given in Fig. 3. The closed circles represent the compositional ratios of the proprietary alloys, whereas the open circles or squares designate the compositional ratios of the experimental alloys. While all the experimental alloys contained 4% palladium, two of them also contained 8%. Each pair of 4 and 8 percent palladium containing alloys is represented by an open square. The range of gold-copper-silver ratios of over 100 Type III and IV gold alloys, as determined by Eich and others (1969), is included for purpose of comparison.

Physical and Mechanical Testing

Specimens for hardness and metallographic tests were cast into plates approximately 2 mm thick. Vickers hardness numbers (VHN) were determined using a diamond pyramid hardness indenter with a one kilogram load on a Kentron Microhardness Tester. Metallographic examination was made with a Zeiss Universal Microscope after appropriate polishing and etching.

X-ray diffraction specimens were prepared by passing magnetically cleaned filings through an 88 micron sieve. All specimens were vacuum sealed in glass tubes and

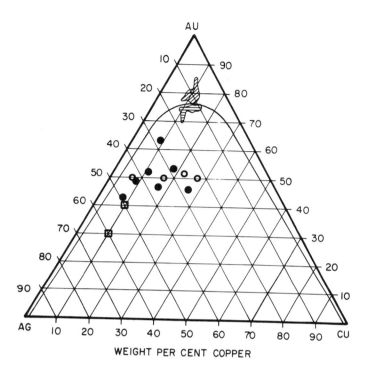

Fig. 3. Gold-silver-copper ratios of experimental and
         proprietary low-gold alloys included in the study.

solution annealed at 700°C for two hours. Heat treatments were conducted at
350-400°C for 30 minutes. X-ray diffraction tests were run at 20 ma and 35 KV
using a Norelco Model 50-100-00 Diffractometer equipped with a fine focusing
copper tube and a focusing graphite monochrometer.

Clinical Testing

In addition to determining heat hardening characteristics and mechanical
properties, all compositions were fabricated as permanent restorations and
inserted into a series of patients. Using intra-oral photography each of the
castings was subsequently evaluated for tarnish and corrosion on an annual basis
for periods of up to three years. To date, approximately 250 castings of
proprietary and experimental compositions have been inserted and evaluated. The
distribution of the various alloys is seen in Table 4.

While generally speaking the alloys were randomly assigned to various intra-oral
locations, at least one control alloy (Modulay-77% gold) was intentionally
inserted into each patient.

TABLE 4 Distribution of Clinical Samples

Proprietary			Experimental	
Alloy	No.		Alloy	No.
*Modulay	48		UNC #1	9
Maxigold	24		UNC #2	8
Mowrey #120	18		UNC #3	10
Midigold	23		UNC #4	9
Midas	19		UNC #5	8
Mowrey #46	15		UNC #6	6
Forticast	17		UNC #7	5
Minigold	20		UNC #8	9
TOTAL	184			64

*Control

RESULTS AND DISCUSSION

Physical and Mechanical Characteristics

All of the reduced gold alloys included in this study were shown to be heat hardenable. When water quenched from solution annealed temperatures, all alloys exhibited minimum hardness. At reduced temperatures the hardness values increased. While the heat hardening response of the reduced gold alloys were similar to conventional gold-based alloys, two basic differences were noted. First, the low-gold alloys reached maximum hardness at 50°C or more above most Type III and IV gold alloys. Secondly, while the latter maintains maximum hardness values at optimum aging temperatures indefinitely, the reduced gold alloys began to overage in 2-3 hours.

Alterations of gold-silver-copper ratios can effect the hardness and other mechanical properties of an alloy. In the solution annealed or softened state, substitution of gold atoms for silver atoms or visa versa results in minimal solution hardening. This is expected since the electrostatic and geometric properties of the atoms are so similar. For example, the interatomic spacing for these two metals differs by less than 0.3%. Indeed, the greatest change of hardness in the softened state results from substitution of copper for gold or silver (Wise, Crowell, and Eash, 1932; Leinfelder and Kusy, 1980). This effect may be demonstrated in Table 5 via the (Au+Ag)/Cu ratio.

The first column includes the different proprietary alloys, followed by their respective (Au+Ag/Cu) ratios. The last three columns indicate the measured mechanical properties: VHN, proportional limit (P.L.), and ultimate tensile strength (U.T.S.). All the properties were determined after solution annealing at 700°C for two hours and quenching. When the conventional high-gold and seven low-gold alloy copper contents were compared, an inverse ranking between the (Au+Ag/Cu) ratios and the mechanical properties were noted. In other words, those alloys with the greatest amount of copper substitutional hardening

TABLE 5 (Au+Ag)/Cu vs. VHN, P.L., & U.T.S.
of Proprietary Alloys (Softened State*)

Alloy	(Au+Ag)/Cu	VHN	P.L.(MPa)**	U.T.S.(MPa)**
Modulay	12.13	131	190	360
Midas	10.69	141	230	390
Midigold	8.45	145	240	390
Maxigold	10.09	151	320	450
Minigold	11.60	165	230	370
Mowrey #46	5.13	172	390	540
Mowrey #120	4.59	189	410	530
Forticast	3.25	212	430	590

*Solution annealed at 700°C for 2 hr and quenched.
**Average of mean values found in Tables 3 & 4 of Kusy and Leinfelder (1980).

demonstrated the highest hardness and tensile values. It should be noted that, in the softened state, the relationship between mechanical properties and the Au/Cu ratios was less definitive than the (Au+Ag/Cu) ratios (Kusy and Leinfelder, 1980). Finally, note that the first five alloys apparently fell into one group, while the last three alloys fell into a second group.

The ranking between composition and properties is further demonstrated in Table 6. The first column presents the various experimental alloys, followed by their respective (Au+Ag/Cu) ratios and microhardness values in the solution annealed state. As in the previous figure, an inverse relationship prevails between the (Au+Ag/Cu) ratios and VHN. Again, note that there is a general ranking of the first five alloys into one group followed by the last three alloys into a second group. Alloy #'s 4 & 7 and 5 & 8 had the same compositions except that one contained 4% palladium and the other 8%. Since the hardness values were not affected dramatically, palladium was not considered in the noble metal/Cu ratios.

TABLE 6 (Au+Ag)/Cu vs. VHN of Experimental Alloys (Softened State*)

Alloy	(Au+Ag)/Cu	VHN
UNC #3	13.33	110
UNC #5	9.00	115
UNC #8	9.00	127
UNC #4	9.00	136
UNC #7	9.00	138
UNC #6	3.55	152
UNC #1	2.71	170
UNC #2	4.71	188

*Solution annealed at 700°C for 2 hr and quenched.

Although precision casting in dentistry can routinely produce tolerances of less than 20 microns, the space between the casting and the adjacent tooth structure (i.e., margins) frequently must be adjusted by as much as 100 microns through a process known as burnishing. After mounting the casting on a die, the laboratory technician or dentist hand rubs a small ball burnisher or other suitable instrument over the surface of the casting. Starting in an area approximately 1.5-2 mm above the margin, the clinician translates the instrument in an oscillating manner under heavy pressure in a direction parallel to the casting margin until the margin is closed. Contrary to popular belief, this technique does not cause a wholesale migration of wrought metal across the casting surface to the margin, but instead produces a deformation of the casting itself.

Concurrent with the proper thickness and design of a casting, at least two mechanical properties control the burnishability of an alloy: They are ductility and yield strength. In general the higher the ductility and the lower the yield strength, the more burnishable is the alloy. This property can be estimated by the following expression:

$$\text{B.I.} = \text{Burnishability Index}$$

$$= \frac{\text{Ductility}}{\text{Yield Strength}} \tag{1}$$

The burnishability index for the proprietary alloys are given in Table 7. All the alloys are listed in order of increasing hardness after solution

TABLE 7 Burnishability Index of Proprietary Alloys

Alloy	B.I. (%/MPa)*	B.I.(%/MPa)**
Modulay	0.231	0.167
Midas	0.109	0.127
Midigold	0.146	0.114
Maxigold	0.097	0.098
Minigold	0.117	0.069
Mowrey #46	(0.051)	0.058
Mowrey #120	(0.063)	0.039
Forticast	0.063	0.023

*Computed from definition of burnishability (cf equation 1).
**Computed from empirical expression (cf equation 4).

treating. The table contains two separate sets of burnishability indices: The first is calculated from the quotient of ductility and yield strength measurements for a specific alloy; the second set is deduced from empirical relationships for low-gold alloys. The derivation of this index can be expressed by the following equations (cf Table 6 and Fig. 7 of Kusy and Leinfelder, 1980):

$$\text{Elongation (\%)} = 287 \, [\exp(-1.6 \times 10^{-2} \, \text{VHN})] \tag{2}$$

$$\text{Proportional Limit (MPa)} = 2.51 \, \text{VHN}-117.13 \tag{3}$$

$$\text{B.I.} = \frac{(\text{equation 2})}{(\text{equation 3})} = (\text{VHN}-46.7)^{-1} \, [114.3 \, \exp(-1.6 \times 10^{-2} \, \text{VHN})] \tag{4}$$

Because elongation is so sensitive to specimen preparation, every effort was made to keep at least the specimen geometry constant. Consequently, the ductility measurements were abstracted from a recently completed study in which manufacturers' dumbbell shaped specimens were evaluated. Since only rod shaped specimens were received from Mowrey, the actual manufacturer's data was substituted (Bolger, 1980). These more ductile values are reflected in the burnishability indices differentiated by parentheses. In contrast, yield strength or proprotional limit values were taken from both dumbbell and straight rod castings, since these properties were relatively insensitive to specimen geometry.

From the values presented in Table 7 the burnishability index for the control alloy, Modulay, is the highest. This is followed by Midas, Midigold, Maxigold, and Minigold. The lowest group contains Mowrey #46, Mowrey #120, and Forticast. Similar rankings were obtained when the (Au+Ag/Cu) ratios were considered (cf Table 5). Interestingly, these rankings concurred with the clinical evaluations.

Given the proper ductility and strength measurements, it should be possible to estimate the "burnishability limit", that is, the minimum acceptable burnishability index which a serviceable alloy can have. For a casting alloy to be readily burnishable, experience dictates that the proportional limit should not exceed approximately 207-276 MPa and that the percent elongation should not be less than 25-30%. On this basis, the burnishability limit (via equation 1) is about 0.091. From Table 8, which equates microhardness to burnishability index (via equation 4), the maximum acceptable VHN should be approximately 153.

TABLE 8   Relationship Between Burnishability Index
and Vickers Hardness for Low Gold Alloys*

VHN	B.I. (%/MPa)
100	0.433
110	0.311
120	0.229
130	0.171
140	0.130
150	0.100
160	0.078
170	0.061
180	0.048
190	0.038
200	0.030
210	0.024
220	0.020

*Computed from empirical expression (cf equation 4).

On the basis of the foregoing discussion it should be possible to determine the relative burnishability of an alloy without determining the proportional limit and percent elongation. Given the VHN measurements (cf Tables 6 & 8), alloy #'s 1 & 2 should be the most difficult to burnish and #'s 3 & 5 should be the easiest. These predictions corroborate the clinical observations.

In the hardened state, order-disorder and precipitation reactions overwhelm any effects from solid solution hardening. Consequently, the Au/Cu ratio is most important (Paffenbarger, Sweeney, and Isaacs, 1932; Kusy and Leinfelder, 1980). As Table 9 shows, an inverse relationship is suggested between the Au/Cu ratios of the proprietary alloys and their mechanical properties. Table 10 shows this same relationship for the experimental alloys in the hardened state.

TABLE 9 Au/Cu vs. VHN, P.L., & U.T.S. of Proprietary Alloys (Hardened State*)

Alloy	Au/Cu	VHN	P.L. (MPa)**	U.T.S. (MPa)**
Modulay	10.27	110	180	360
Maxigold	7.00	228	400	600
Midas	5.75	229	470	630
Minigold	5.33	243	410	620
Midigold	4.95	245	520	660
Mowrey #46	2.80	277	630	810
Mowrey #120	2.94	284	680	780
Forticast	2.01	299	590	850

*Age-hardened at 400°C for 30 minutes.
**Average of mean values found in Tables 3 & 4 of Kusy and Leinfelder (1980).

TABLE 10 Au/Cu vs. VHN of Experimental Alloys (Hardened State*)

Alloy	Au/Cu	VHN
UNC #3	7.16	180
UNC #1	1.85	210
UNC #4	4.00	219
UNC #7	4.00	223
UNC #8	3.00	237
UNC #5	3.00	239
UNC #2	2.86	275
UNC #6	2.36	287

*Age-hardened at 350°C for 30 minutes.

Corrosion Resistance

After twelve months of service, the low gold alloys generally showed no evidence of tarnish or corrosion. Sometime within the second year, however, some proprietary alloys started to tarnish in interproximal areas, particularly near the gingival crest. While the tarnish or corrosion products normally appeared in localized areas, at least one casting exhibited discoloration over an extensive area. A typical example of the localized discoloration product is illustrated in Fig. 4. Note that it occurred near the gingival ridge where the plaque build-up was highest and the electrolyte cell concentration was greatest. The more general, but seldom seen, corrosion state is shown in Fig. 5. Additionally, small discolored patches were observed on some of the experimental compositions in which the gold content ranged from 28 to 39 weight percent. Electron microprobe analysis revealed that all surface deposits were silver sulfide. None of those compositions containing more than 42.0 percent gold gave any evidence of tarnish or corrosion over a three year period.

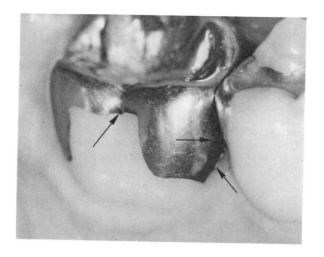

Fig. 4. Typical example of localized tarnish.

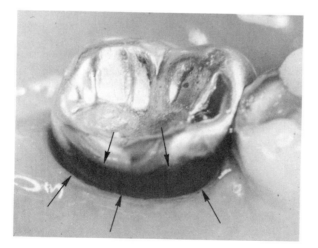

Fig. 5. Example of generalized type of corrosion.

Microstructure

In the solution annealed state, all alloys exhibited a uniform equiaxed grain structure. Other than occasional twinning, which is normally associated with gold-based alloys, all compositions had a single-phase grain structure. When the experimental and proprietary alloys were aged for 30 minutes at 350°C and 400°C, respectively, no microstructural changes were observed at magnifications up to 1000 diameters. When the alloys were overaged, either by heat treating at elevated temperatures or by prolonging the heat treatment time, significant microstructural modifications occurred. After etching with aqua regia, iodine, or ammonium persulfate cyanide etchants, most compositions revealed that a discontinuous grain boundary precipitate was present. Figure 6 shows a representative microstructure of the alloys in this overaged condition. This microstructure is typical of overaged high gold alloys, particularly those of Type IV composition (cf Fig. 6 of Leinfelder and Taylor, 1977). In each case, the grain boundary precipitate consists of a lamellar two-phase structure. Such an intergranular structure apparently results from a partial separation of the alloys into silver-rich and copper-rich equilibrium phases. This observation is expected, since all alloy compositions fall into the two-phase region of the solubility loop of the gold-silver-copper ternary when heated to 350-400°C.

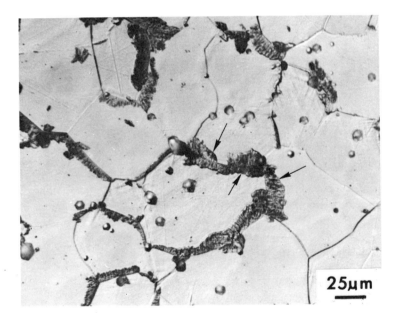

25μm

Fig. 6.   Microstructure of low-gold alloy in over-aged state. The
          discontinuous precipitation occurring at the grain
          boundaries consists of a lamellar two-phase structure.

X-ray diffraction

In the solution treated state, all proprietary alloys exhibited a face-centered cubic structure. After aging at 350°C for short periods of time, however, dramatic changes appeared in the (111) region of their diffraction patterns. Essentially the original (111) line, representing the gold-silver-copper matrix phase, began to decompose - giving rise to two additional peaks. As predicted by the ternary phase diagram, these developing peaks were identified as the (111) silver- and copper-rich phases. As determined on separate but identically treated samples, the rate of hardness increase paralleled the decomposition rate of the original matrix phase into two end phases. In addition to the decomposition described above, ordering was noted. From the diffraction angle of the ordered line, only the copper-rich, AuCu phase ordered. A similar analysis is underway for the experimental alloys.

CONCLUSIONS

Low-gold alloys may be suitable substitutes for conventional gold casting alloys if two conditions are met. First, in order to prevent the occurrence of tarnish and corrosion, the absolute gold content must be at least 42 weight percent. Such low-gold contents can be accommodated only if the palladium content is at least three to four percent. Secondly, in order to ensure that an alloy is readily burnishable, the gold plus silver to copper ratio should be at least 10:1. Increasing the copper content at the expense of gold and silver only reduces ductility and elevates hardness, i.e., decreases the burnishability index.

Since the low-gold alloys discussed herein are essentially an extension of the conventional casting alloys into the two-phase region of the gold - silver - copper diagram, it is not surprising that their microstructure and solid state transformations are similar.

ACKNOWLEDGEMENT

This investigation was supported by NIH research grant Nos. DE-04101, DE-02668, RCDA No. 00052 and RR-05333.

REFERENCES

American Dental Association Specification No. 5 for Dental Casting Gold Alloy, (Approved April 1965. Effective April 1, 1966.) 146-149.
Bolger, G. (1980). Private communication.
Eich, J.D., H.J. Caul, T. Hegdahl, and G. Dickson (1969). Chemical composition of dental gold casting alloy and dental wrought gold alloys. J. Dent. Res., 48, 1284-1289.
Kusy, R.P., and K.F. Leinfelder (1980). Age-hardening and tensile properties of low gold (10-14 kt.) alloys. J. Biomed. Mat. Res. (submitted for publication).
Leinfelder, K.F., and R.P. Kusy (1980). Current status of low-gold alloys. N.Y. State Dent. J. (in press).
Leinfelder, K.F., and D.F. Taylor (1977). Hardening of gold-based alloys. J. Dent. Res., 56, 335-345.
Paffenbarger, G.C., W.T. Sweeney, and A. Isaacs (1932). Wrought gold wire alloys: Physical properties and a specification. J. Amer. Dent. Assoc., 19, 2061-2086.
Report of Conference on Dental Inlay Gold Alloys at National Bureau of Standards (Abstracted Report), March, 1931.
Wise, E.M., W.S. Crowell, and J.T. Eash (1932). The role of the platinum metals in dental alloys. Trans. A.I.M.E., 99, 363-412.

Consideration of Some Factors Influencing Compatibility of Dental Porcelains and Alloys, Part I: Thermo-Physical Properties

R. P. Whitlock*, J. A. Tesk**, G. E. O. Widera***, A. Holmes**,
and E. E. Parry****

*Research Associate, Capt., U. S. Navy, National Naval Dental Center
**Dental and Medical Materials Group, Polymer Science and Standards Division
****Research Associate, DTC, U.S. Navy, National Naval Dental Center
National Bureau of Standards, Washington, DC 20234
***University of Illinois, Materials Engineering Department
Chicago Circle, Chicago, IL 60611

ABSTRACT

Preliminary results on the determination of thermal expansion of several brands of dental porcelain and veneering alloys are presented.[1] The porcelain specimens were cooled rapidly, similar to processing of porcelain veneered alloy prostheses. Large differences in expansion of up to ∿ 30 percent were found to exist between some porcelains and alloys. These differences would virtually preclude the use of those combinations if expansion alone determined compatibility. Varying amounts of trapped excess volume were recovered during heating of porcelain expansion specimens. The trapping of this excess volume in rapidly cooled porcelain changes the thermal expansion relative to that for annealed porcelain. The proposed thermal stress compatibility index of Ringle and others (1978, 1979) and Fairhurst, Harshinger, and Twiggs (1980) should employ the thermal expansion of rapidly cooled porcelain.

KEYWORDS

Ceramics; dental alloys; dental materials; dental porcelain; glass to metal seals; gold alloys; nickel alloys; nickel chromium; porcelain/metal compatibility; thermal expansion, alloys; thermal expansion, glasses; thermal expansion, porcelain.

INTRODUCTION

The increase in the price of gold over the past ten years has resulted in a corresponding increased use of non-precious and low gold alloys in dentistry. In particular, the substitution of non-precious metals for gold alloys has been highest for porcelain fused to metal restorations. Numerous non-precious alloys exist on the market, with new ones appearing rapidly. These new alloys are

---

[1]The use of brand names in this publication is solely for the purpose of rendering the information contained more useful and implies no endorsement of said products by the National Bureau of Standards or Department of the Navy nor is such listing to be used as an indication of efficacy of said products.

designed by their manufacturers to be "compatible" with one or more of the porcelains on the market. The latter were originally designed to function as veneering materials for gold alloys. However, the chemical, mechanical, and physical properties of the new non-precious alloys differ markedly from those of the gold alloys which they replace. Therefore, new techniques are required for successful fabrication of prostheses using these alloys with dental porcelains. Such processing has generally been accepted as being more sensitive to deviations from the prescribed technique when compared with processing of the more traditional noble metal compositions. The differences in properties and techniques when coupled with the variable characteristics of ceramic veneering materials can lead to unpredictable servicability in porcelain-fused-to-metal (PFM) restorations.

A major source of failure in PFM restoratons involves cracking or spalling of the veneered porcelain. These problems are thought to be primarily associated with residual stresses resulting from differences between the thermo-physical and mechanical properties of the ceramic and metal components as well as geometrical factors (Fig. 1) imposed by the design of the tooth preparations, etc. These stresses will vary from one combination of porcelain and alloy to another. They will depend in part on the particular processing techniques which are employed. Because the use of non-precious alloys in dentistry can produce significant reductions in the costs of finished crowns, the objective of this investigation was to determine the properties of porcelains and alloys as an aid in determining conditions which produce the most compatible[2] combinations.

Factors Which Influence the Development of Residual Stress in Porcelain Veneers

a) **Thermal Expansion**
b) **Tg—The Glass Transition Temperature**
c) **Trapped Residual Stress (Strain)**
d) **Temperature Dependence of Elastic Modulus**
f) **Crystallinity**
g) **Thermal Conductivity** ⎱ **Thermal**
h) **Heat Capacity** ⎰ **Diffusivity**
i) **Number of Layered Components**
j) **Relative Layer Thicknesses**
k) **Ability to Relieve Stress**

Fig. 1.   Factors affecting residual stress in dental porcelain fused to metal crowns.

---

[2]For this paper, compatibility is taken to mean stress compatibility.

MATERIALS AND METHODS

Porcelains and alloys chosen for this study are those available from leading manufacturers or involve alloys which are considered by at least one or more of the investigators to be particularly useful for investigation.[3,4]

To accomplish the cited objectives a multitiered approach has been taken. The first phase of the study reported in this paper involved the determination of the individual thermal expansion characteristics of porcelain and alloy specimens which had previously been subjected to either 1, 3, 5, or 7 firing schedules according to manufacturers' recommendations. The equipment and techniques used have been described elsewhere (Whitlock, and others, 1980). (It is considered most important that these determinations are carried out in the same laboratory and with the same equipment, in order to reduce differences which could arise due to systematic error.) In addition to the experimental determinations of thermal expansion of independent porcelain and alloy specimens, a composite, split-ring (Fig. 2) has been used as a device to aid the evaluation of residual stress in a system comprised of alloy with opaque and body porcelains. This is described in a companion paper which follows this one.

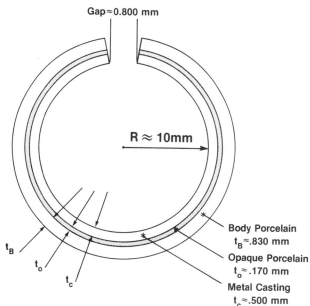

Gap ≈ 0.800 mm

R ≈ 10mm

t_B

t_o

t_c

Body Porcelain
$t_B \approx .830$ mm

Opaque Porcelain
$t_o \approx .170$ mm

Metal Casting
$t_c \approx .500$ mm

Fig. 2   Schematic of porcelain veneered split metal ring

[3]Porcelain materials investigated
    BIOBOND, Dentsply Int., York, PA; CERAMCO, Ceramco Inc., East Windsor, NJ; MICRO-BOND, Howmedica Inc., Dental Division, Chicago, IL; NEYDIUM, J. M. Ney Co., Bloomfield, CT; VITA (VMK-68), Vita Zahnfabrik, H. Rauter GMBH & Co. KG, Bad S'A'ckingen, Germany; WILL-CERAM, Williams Gold, Buffalo, NY

[4]Alloys investigated
    PENTILLIUM; BIOBOND C&B, Dentsply Int., York, PA; CERAMALLOY II, Johnson & Johnson Dental Products, East Windsor, NJ; NP-2, Howmedica Inc., Dental Div., Chicago, IL; JELENKO"O", J. F. Jelenko and Co., New Rochelle, NY; WILLIAMS-Y, Williams Gold, Buffalo, NY; SMG-2, J. M. Ney, Co., Bloomfield, CT

RESULTS

At the present time the thermal expansion of each manufacturers' porcelain listed[3] has been determined at least once for each firing schedule. Some preliminary results are shown in Figs. 3 and 4. Ceramco, Vita, Micro-bond, and Biobond opaque porcelain curves are cumulative results of up to five separate runs, differences between these porcelains are significant at the 95 percent confidence level at 500 °C. The Will-ceram and Neydium porcelain results are from only one run and further testing is required on these opaque porcelains. Precision, however, is expected to be similar for all porcelain products evaluated. Figure 5 depicts the differences which exist between a number of alloys, the precision for each of these curves lies between one and five percent.

Expansion of 1st Fire Opaque Porcelains

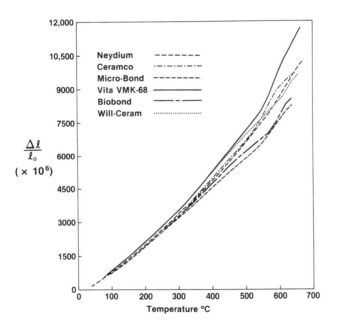

Fig. 3   Typical thermal expansion curves for
         opaque porcelains. Ceramco, Vita,
         Micro-Bond, and Biobond curves are
         the average of multiple determinations.

Table 1 lists preliminary results of the coefficient of thermal expansion values for the six porcelain products investigated, as well as the corresponding values for the alloys. The porcelain expansion values are shown as functions of the number of firing schedules. Alloy coefficients are grouped according to the non-precious alloys (the upper section of the alloy chart) and precious metals (the lower section). No significant differences were found when comparing the number of firing cycles for alloys.

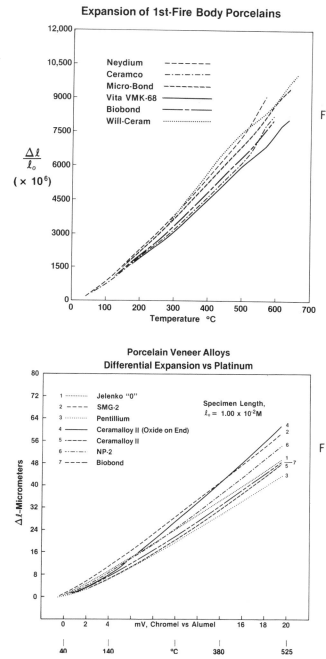

Expansion of 1st-Fire Body Porcelains

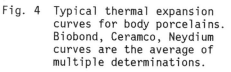

Fig. 4 Typical thermal expansion curves for body porcelains. Biobond, Ceramco, Neydium curves are the average of multiple determinations.

Fig. 5 Thermal expansion of porcelain veneering alloys, Pentillium, Caramalloy II, NP-2, and Biobond. Curves for Ceramalloy II illustrate the importance of oxide removal from specimen faces; the specimen with oxide on the end in contact with pushrods shows approximately 30 percent higher expansion.

TABLE 1 Coefficient of thermal expansion, α, at 500 °C for porcelains and alloys evaluated. Porcelain results are shown as a function of the number of laboratory firing cycles for specimens, prior to expansion determination. The first four alloys, Pentillium, Biobond, Ceramalloy II and NP-2, are non-precious alloys.

| PORCELAIN | No. of Firing Cycles | | | | | | | |
| | OPAQUE | | | | BODY | | | |
	1	3	5	7	1	3	5	7
Biobond	13.10	13.10	13.74	14.35	13.59	14.62	14.29	14.82
Ceramco	14.09	12.92	12.70	14.23	13.02	13.07	12.93	13.05
Microbond	12.38	13.29	13.92	14.80	14.59	14.94	14.68	15.28
Neydium	13.97	13.25	13.45	14.23	15.47	14.04	14.92	15.93
Vita (VMK-68)	15.05	14.29	15.82	16.03	12.73	13.17	13.55	14.45
Will-Ceram	14.68	14.76	14.74	15.45	15.97	16.23	15.94	15.99

ALLOYS	
Pentillium*	14.14
Biobond C&B*	14.27
Ceramalloy II*	14.63
NP-2*	15.28
Jelenko-O**	14.71
Williams—Y**	14.83
SMG-2**	15.73

Thermal Expansion Coefficients ($\alpha\,°C^{-1} \times 10^{6}$) for Temperature Range 40°C—500°C

Comparison of Coefficient of Expansion—Porcelains & Alloys

*Nonprecious Alloy
**Precious Alloy

In all instances, there is wide variation in expansion within groups and between groups. As previously stated, statistically significant differences have been determined to exist between some manufacturers' porcelains, and with the expectation that similar magnitudes of confidence will exist for all determinations, approximately a 20 percent difference exists between the extremes for all porcelains. The study of Neilson and Tucilleo (1972) shows that an approximate seven percent difference between alloy and porcelain expansion could lead to unacceptable stress levels. Therefore, it would be impossible for some porcelain-alloy combinations to function without mitigating factors (for example, viscous flow and/or stress relief of porcelain at lower temperatures).

A close examination of the expansion curves of the alloys, in Fig. 5 reveals that some of the alloy curves cross at lower temperatures; this would change the relative intensity of expansion (contraction) induced residual stress within a porcelain metal system (dental restoration). Hence, the magnitude and sign of stress developed in any combination of porcelain and alloy will depend strongly on the temperature at which the porcelain behaves as a solid.

A schematic of a typical plot of the expansion characteristics of porcelain veneering materials is shown in Fig. 6. The shapes of the heating (upper) and cooling (lower) portions of the curve are different, the cooling curve is characteristic of annealed ceramics. The heating and cooling rates which are employed to obtain such a curve are slow and necessarily differ from the rapid cooling which is used for commercial (dental laboratory) firing schedules. The necessarily slower rates used during measurement procedures appear to effectively anneal the porcelain materials, therefore, their physical and mechanical properties must be altered. This phenomenon of annealing has provided an opportunity to observe a most important effect, excess volume is retained to a varying degree in each porcelain material when cooling rates follow those prescribed by the manufacturers. The relative amount of this residual volume is indicated by the decreased rate of expansion which occurs upon heating and is evaluated during the thermal expansion <u>measurement</u> process, as is illustrated in Fig. 7. In Fig. 7 are shown hypothetical thermal expansion measurement curves derived from using (1) a specimen rapidly cooled from its fired state, as in dental laboratory practice, and (2) an equivalent specimen, slowly cooled prior to measurement. Volume recovery during measurement of the fast cooled specimen is indicated by the decreased rate of expansion relative to the extrapolated (dashed) line. The recovered fractional volume $\Delta v/v$ is approximately three times for fractional dimensional recovery, δ, and

$$\delta \approx \left[\left(\frac{\Delta \ell}{\ell_o} \right)_{ext,\ T_3} - \left(\frac{\Delta \ell}{\ell_o} \right)_{obs,\ T_3} \right]$$

where "ext" refers to the extrapolated curve and "obs" refers to the observed curve.

THERMAL EXPANSION OF GLASS

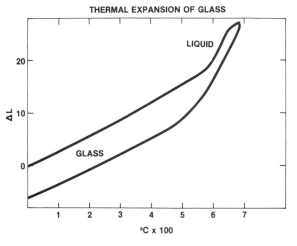

Fig. 6 Typical thermal expansion and contraction curve. The upper section is expansion of a quenched material, heated slowly, the lower curve is from slow cooling. The cooling curve is of the type expected for an annealed specimen.

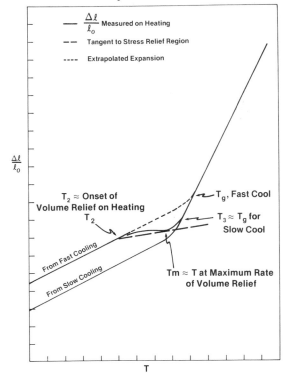

Fig. 7   Hypothetical expansion/contraction curves
for a glassy material. The upper curve,
including the short dashed portion is for
fast cooling, and results in a higher
glass transition temperature, T_g, and some
trapped excess volume, indicated by the
difference between $\Delta \ell/\ell_0$ fast cool, vs
$\Delta \ell/\ell_0$, slow cool. Upon heating of a fast
cooled specimen, recovery of excess
volume occurs, between temperatures T_2
and T_3.

An alternate to this method of analysis for excess volume involves estimating the
differences in $\Delta \ell/\ell_0$ at the temperature, T_m, of maximum rate of volume recovery, as
indicated in Fig. 7.

The derived relative volume recovery values for a variety of ceramic materials,
exposed to one through three firing schedules are shown in Fig. 8 for the opaque
porcelains. For the most part the two methods of analysis give similar results
and the same conclusions are reached from either. There appears to be relatively
little excess volume trapped in opaque porcelains to the left of the figure, with
much larger amounts developed in those on the right. The same comparison for the
body materials is shown in Fig. 9 and again large differences are observed between
porcelains.

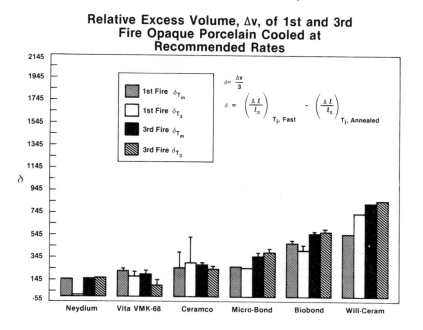

Fig. 8 Relative relief of excess volume in opaque porcelain as indicated by recovery of excess volume during heating of specimen. δ is determined at either of two temperatures, T_m or T_3. (Reference Fig. 7).

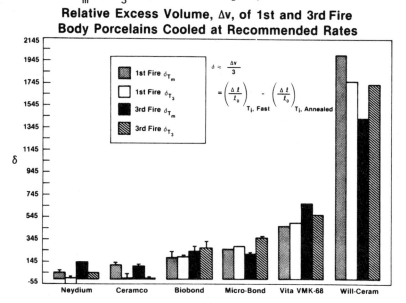

Fig. 9 Relative relief of excess volume in body porcelains indicated by recovery of excess volume during heating of specimen. δ is determined at either of two temperatures, T_m or T_3. (Reference Fig. 7).

CONCLUSIONS

(1)   Significant differences exist in the thermal expansion of dental porcelains
       as a function of temperature.

(2)   An approximate 20 percent difference in extremes of porcelain expansion up
       to 500 °C clearly reveals the potential for much less compatibility between
       some combinations of porcelain and alloy if a range of only ∿ 7 percent is
       acceptable for any given alloy, as calculated by Nielson and Tuccillo (1972).

(3)   Varying amounts of excess volume are trapped within different porcelains which
       are rapidly cooled according to laboratory processing conditions.

(4)   Trapping of excess volume dictates the use of thermal expansion data from
       rapidly cooled, unannealed specimens in evaluation of the compatibility
       index proposed by Ringle and others (1978) and modified by Fairhurst
       and others (1980).

(5)   Due to the recovery characteristics of some porcelains, heat treatment of a
       porcelain veneered crown at a temperature above the strain point, T_2, near
       T_m may be beneficial; one possibility is isothermal annealing near T_m.

REFERENCES

Fairhurst, G. W., D. Harshinger, and S. W. Twiggs (1980). Glass transition
    temperatures, T_g, of porcelain. J. Dent. Res., 59A, Abstract 656.
Neilson, J. P. and J. J. Tuccillo (1972). Calculation of interfacial stress
    in dental porcelain bonded to gold alloy substrate. J. Dent. Res., 51,
    1043-1047.
Ringle, R. D., D. T. Harshinger, K. J. Anusavice, and G. W. Fairhurst (1979).
    Thermal contraction behavior of alloy-opaque porcelain body porcelain
    systems. J. Dent. Res., 58A, Abstract 606.
Ringle, R. D., R. L. Weber, K. J. Anusavice, and C. W. Fairhurst (1978).
    Thermal expansion/contraction behavior of dental porcelain-alloy systems.
    J. Dent. Res., 57A, Abstract 877.
Whitlock, R. P., J. A. Tesk, E. E. Parry, and G. Dickson (1980). Observations of
    significant differences in thermal expansion characteristics of dental
    porcelains. J. Dent. Res., 59B, Abstract No. 40.

[5]This investigation was supported by the Naval Medical Research and Development
Command and by the National Institutes of Health, National Institute of Dental
Research, Interagency Agreement No. YO1-DE-40015.

Consideration of Some Factors Influencing Compatibility of Dental Porcelains and Alloys, Part II: Porcelain/Alloy Stress

J. A. Tesk*, R. P. Whitlock**, G. E. O. Widera***,
A. Holmes*, and E. E. Parry****

*Dental and Medical Materials Group, Polymer Science and Standards Division
**Research Associate, Capt., U.S. Navy, National Naval Dental Center
****Research Associate, DTC, U.S. Navy, National Naval Dental Center
National Bureau of Standards, Washington, DC 20234
***University of Illinois, Materials Engineering Department
Chicago Circle, Chicago, IL 60611

ABSTRACT

The change in gap of a porcelain veneered split metal ring has been previously proposed as an indication of compatibility (Whitlock and others, 1980). Gap changes were measured for several combinations of opaque and body porcelain fired on a dental alloy ring. Finite element and theoretical calculations of the gap change, when compared with experimental observations, indicate stress relief in porcelain-metal composite systems at temperatures well below the glass transition temperature, T_g. Extension of the bimaterial compatibility index to more complex (more than two layer) systems is shown to require further consideration. Stress analyses based on Timoshenko's equation for shape change of multimaterial thermostats leads to oversimplified states of stress and cannot yield the detailed information on shear and tensile stresses needed for viscous flow above T_g and brittle fracture below T_g.

KEYWORDS

Bonding, glass to metal; dental alloys; dental porcelain; finite element analysis, porcelain on metal; glass to metal seals; porcelain/alloy compatibility; porcelain/alloy composite structures; split ring compatibility test; stress analysis glasses; stress in dental alloys.

INTRODUCTION

This study involves the change in gap of porcelain veneered split metal rings.(Fig. 1) The gap changes are to be used as indicators of stress within the composite systems. The smaller changes are considered indicative of smaller overall states of stress.

---

[1]The use of brand names in this publication is solely for the purpose of rendering the information contained more useful and implies no endorsement of said products by the National Bureau of Standards or Department of the Navy nor is such listing to be used as an indication of efficacy of said products.

Experimental measurements are to be correlated with theoretical calculations of the change in gap. From this, detailed information on the stress within any system can be analyzed.

MATERIALS AND METHODS

Approximate dimensions for metal, ring, opaque and body porcelain layers, and the size of the gap were shown in Fig. 2 of the preceeding paper, Whitlock and others (1980). The rings are formed from polyethylene plastic patterns, which are then sprued, invested, burned out, and cast with the desired alloy. After the casting process, the rings are machined to the predetermined size and split. Following the alloy manufacturer's recommendations, the metal is conditioned for application of opaque porcelain according to dental laboratory procedures. The initial gap dimensions at a, b, and c (Fig. 2) and four diameterial measurements are taken as indicated in Fig. 3, with the aid of a traveling microscope. The opaque layer is applied and machined in the green state to a uniform thickness using a quartz bit on a lathe and then fired. The process is repeated for the body application. The precision with which the various thicknesses of layers can be controlled is depicted in Table 1. A ring with body porcelain prior to firing is shown in Fig. 1. After each firing, dimensions are rechecked. Also checked are the thicknesses of the layers at 15 positions around the ring to assure uniformity of the coating and obtain average thicknesses for use in theoretical calculations. The presence of a gap itself may have an effect on the overall response of the ring to any particular combination of materials, i.e., edge effects may change the shape of the ring to something other than circular. Therefore, the diameter of the ring is also measured at several positions, (Fig. 3), following each firing. Three rings of each combination were investigated.

TABLE 1   Typical thicknesses of porcelain layers and metal ring showing consistency to approximately two percent.

SPLIT RING COMPATIBILITY TEST

	CERAMCO	VITA
METAL	0.511±.004	0.512±.007
OPAQUE	0.668±.026	0.685±.015
BODY	1.538±.020	1.510±.018

Fired Mean Total Thickness (mm)

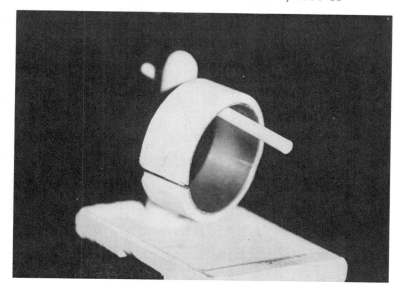

Fig. 1 Split ring with body porcelain trimmed to
proper thickness, just before firing.

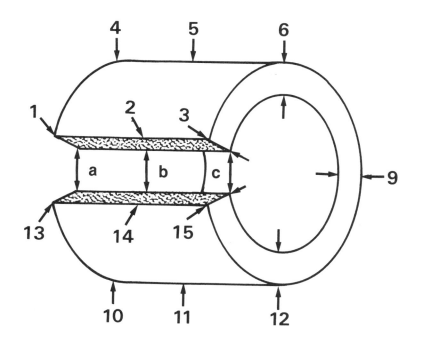

Fig. 2 Schematic showing positions a, b, and c which gap
dimensions are measured and the 15 locations used
to arrive at average thicknesses for the metal ring
and each layer of porcelain.

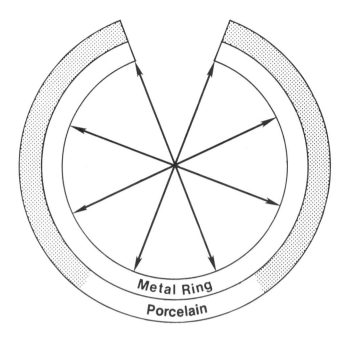

Fig. 3 Schematic of ring, showing the four diametrical
        measurements employed to check overall shape of
        ring after each firing.

To correlate the experimentally determined porcelain expansion values with the
change in gap between measurements taken before and after firing and to help
determine the magnitude and location of residual stresses, finite element models
(FEM) of the split ring are being constructed. A schematic of one model (Fig. 4)
shows the layers (not all need be used) which are divided into a number of small
nearly rectangular elements, connected at their corners (nodes). Near the edges of
the gap smaller elements are used so that the influence of edge effects will be
reflected. The split ring employed in this study holds promise as a sensitive
method for determination of compatibility, under the assumption that minimal change
in gap means minimal overall stress. Experimental results on changes in gap for
rings coated with two brands of porcelain, Vita-VMK68, (opaque plus body) and
Ceramco, (opaque plus body), are shown in Table 2 the alloy is Pentillium.

As was shown in Fig. 3 of the preceeding paper (Whitlock and others, 1980) the
expansion of the Vita opaque porcelain is generally quite different from that of the
Ceramco opaque, and from Fig. 4 (ibid.) the expansion of the Vita body porcelain
differed by only a few percent from that of Ceramco body porcelain. Due to stronger
effects of the larger, outer layer on shape change, little difference in behavior
between Vita or Ceramco veneers would be expected on the basis of thermal expansion
alone. Overall, however, the Vita veneer produced a net change of -16 percent in
gap as opposed to -2.5 percent for the Ceramco veneered ring, i.e., the closure was
more than 500 percent greater. In Fig. 5 are also shown the net preliminary results
for Will-Ceram and Biobond porcelains, which show <u>increases</u> in gap as opposed to
the gap closures with Vita and Ceramco.

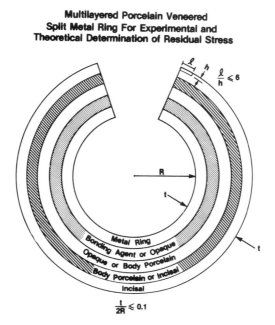

Fig. 4 Schematic of first finite element model FEM. More elements are used near gap to more precisely obtain effects of edges on overall shape. The ratio $\ell/h \leq 5$ is maintained, and the overall thickness to diameter ratio, $t/2R \lesssim 0.1$ is used to simplify theoretical approaches.

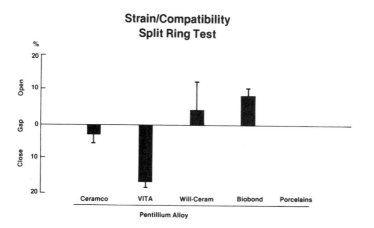

Fig. 5 Net opening or closing of gap for split rings of Pentillium alloy, with veneers of several opaque and body porcelains fired according to porcelain manufacturers recommended schedules.

The standard error in reproducibility of the gap measurements in three of the porcelain/alloy combinations shown is two to three percent, whereas, for the Williams-Pentillium combination it is 13 percent. This may indicate that the latter combination might produce less consistent results in the dental laboratory. (Tests will be repeated to verify this behavior.)

No simple correlation was found between the expansion of the porcelain (Whitlock and others, 1980) at the glass transition temperature T_g, and split ring gap change. (T_g was determined by extrapolation of the expansion curves from low temperature to intersection with the high temperature region of expansion.) Shown in Table 2 is a comparison between experimental results of the change in gap, and those predicted by calculations using the expansion values previously determined. The calculations, based on the Timoshenko (1925) analysis for bimaterial strips, assume a constant curvature. The two layer calculation was made using a linear weighted average of the expansion values of third fire opaque and second fire body porcelains. The FEM predicts gap changes similar to the analytic solutions. The differences between experimental and theoretical values indicate that stress relief of some kind is occurring below T_g, resulting in smaller than calculated dimensional changes in the ring. This conclusion is in agreement with calculations by Bertolotti (1980) and Cascone (1979). The FEM also predicts the development of an elliptical shape, in agreement with the experimentally determined shape after firing, Fig. 6.

The value of FEM is further indicated in Fig. 7, which describes the fundamentals involved in a change in shape, based on the original Timoshenko (1925) analysis of a two body, or bimaterial strip. If two pieces of material, with different coefficients of expansion are heated they expand to different lengths. To make the lengths the same, the upper one must be pulled, or put in tension, with compression applied to the lower one. To make them retain dimensions after joining, equilibrium conditions require that a bending moment be imposed. This puts the system into a state which produces curvature. These simple stress systems imposed by Timoshenko were used to estimate shape changes for use in design of thermostats. However, the reverse procedure, i.e., measuring changes in shape and applying that simple analysis for determining stress can only lead back to oversimplified states of stress and does not yield the detailed information needed for determination of compatibility. For example, a discontinuity exists in calculated stress at the interface as shown in the block in Fig. 7. This arises from the simple stress analysis. Clearly, shear must also be present at the interface and this could be of particular importance for stress relief which may occur during cooling while the porecelain is in the viscous state. Hence, the Timoshenko (1925) two-layer or the Savolainen-Sears (1969) three-layer analyses, developed to predict shape or dimensional changes, are incapable of providing the detailed information needed regarding stress. For this reason, FEM are being used to correlate gap and shape changes of the split metal ring with expansion and stress-relief behavior of the system. The results of FEM (Widera, Tesk, Whitlock, 1980), for the porcelain on metal rings evaluated thus far show that, indeed, high shear-stresses exist at boundaries between layers and that the thinner the layer and the larger the differences in expansion across the layer, the larger the shear stresses along the interface. Large tensile stresses must also exist, and these must be of prime importance for consideration of internal failure mechanisms at lower temperatures involving brittle porcelain layers and interfaces.

TABLE 2   Theoretical and experimental changes in gap
          before and after firing of porcelain veneered
          split metal rings. Dimensions are expressed
          in mm. The lower values obtained experimentally
          indicate stress relief in porcelain at temper-
          atures lower than the glass transition temper-
          ature, T_g.

PORCELAIN ALLOY	VITA MK 68 PENTILLIUM	CERAMCO PENTILLIUM
Experiment	-0.133	-0.024
THEORETICAL MODELS		
Timoshenko (1925) 2-layer analysis	-0.275	-0.166
Timoshenko and Savolaine-Sears (1969) 3-layer analysis	-0.378	-0.304
FEM 3-layer ring	-0.394	-0.234

FEM Model Will Be Used to Check Departures From Uniform Concave Curvature.

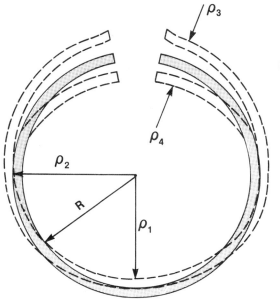

Fig. 6 Edge effects produce non-uniform
       stress resulting in development
       of non-uniform curvature after
       firing of porcelain veneered
       split metal ring. $\rho_i \neq R$, where
       R is the curvature of the ring
       before firing.

J.A. Tesk

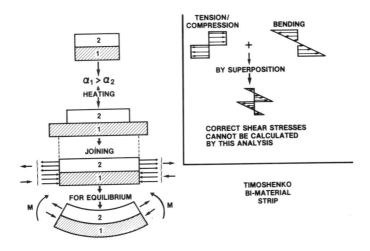

Fig. 7 Fundamentals of Timoshenko analysis for shape
        change (curvature) of bimaterial strip. The
        stress discontinuity shown in the boxed section
        demonstrates that shear stress must be present
        along the interface; this cannot be calculated
        from this analysis as only pure bending and
        tensile/compressive stresses along the interface
        are allowed, by assumption.

CONCLUSIONS

(1) Large amounts of stress relief, in the multilayered systems comprising
     dental restorations, apparently take place at temperatures well below the
     porcelain glass transition temperature.

(2) A split ring, with the use of FEM holds promise for revealing information on
     the stress compatibility of porcelain metal systems, and the split ring
     alone may prove to be a useful, simplified tool for quality control and
     evaluation of compatibility.

(3) A simple compatibility index as proposed by Ringle and others (1978, 1979)
     based on stress analysis obtained from a Timoshenko bimaterial strip is a
     useful concept for two layer considerations and as a first approximation for
     stress but should be used with caution when more complex systems of three or
     more layers are considered due to the complex stresses developed.

REFERENCES

Bertolotti, R. L. (1980). Generalized theory of porcelain-metal and porcelain-porcelain compatibility. J. Dent. Res., 59A, Abstract 662.

Cascone P. J. (1979). Effect of thermal properties on porcelain-to-Metal compatibility. J. Dent. Res., 58B, Abstract 683.

Neilson, J. P. and J. J. Tuccillo (1972). Calculation of interfacial stress in dental porcelain bonded to gold alloy substrate. J. Dent. Res. 51, 1043-1047.

Ringle, R. D., D. T. Harshinger, K. J. Anusavice, and G. W. Fairhurst (1979). Thermal contraction behavior of alloy-opaque porcelain body porcelain systems. J. Dent. Res., 58A, Abstract 606.

Ringle, R. D., R. L. Weber, K. J. Anusavice, and C. W. Fairhurst (1978). Thermal expansion/contraction behavior of dental porcelain-alloy systems. J. Dent. Res., 57A, Abstract 877.

Savolainen, U. U. and R. M. Sears (1969). Thermostat metals. In A. E. H. Dietz (Ed.), Composite Engineering Laminates, Chapter 10, MIT Press, Cambridge, MA.

Timoshenko, S. (1925). Analysis of biometal thermostats. J. Opt. Soc. Amer., 59A, 233-255.

Whitlock, R. P., J. A. Tesk, E. E. Parry and G. E. O. Widera (1980). A porcelain veneered split metal ring for evaluation of compatibility of dental porcelain-alloy systems. J. Dent. Res., 59A, Abstract 682.

Whitlock, R. P., J. A. Tesk, G. E. O. Widera, A. Holmes, and E. E. Parry (1980). Consideration of some factors influencing compatibility of dental porcelains and alloys, Part I: Thermophysical properties, Proceedings of 4th Annual Meeting, International Precious Metals Institute, Toronto, Ontario, Canada, Pergamon Press.

Widera, G. E. O., J. A. Tesk, and R. P. Whitlock (1980). Theoretical stress analysis of porcelain veneered split metal ring for evaluation of compatibility of porcelain alloy systems. J. Dent. Res., 59A, Abstract 682.

---

[5]This investigation was supported by the Naval Medical Research and Development Command and by the National Institutes of Health, National Institute of Dental Research, Interagency Agreement No. YO1-DE-40015.

Corrosion and Tarnish of Ag-In and Ag-In-Pd Alloys

T.K. Vaidyanathan* and A. Prasad**

*Associate Professor, Department of Dental Materials Science, New York University Dental Center, 345 East 24th Street, New York, N.Y. 10010
**Director of Research and Development, Jeneric Industries Inc., 125 North Plains Industrial Road, Wallingford, Connecticut, 06492

ABSTRACT

This investigation reports the results of anodic potentiodynamic polarization in 1% sodium chloride solution and the alternate immersion and wiping tests in 0.5% Na_2S solution. Ag-In binary and Ag-In-PD ternary alloys were studied. In alloying does not significantly decrease the corrosion or tarnish resistance of Ag while Pd alloying does.

KEYWORDS

Anodic polarization; potentiodynamic polarization; alternate immersion; Pd alloying; In alloying; corrosion resistance; tarnish resistance.

INTRODUCTION

In recent years, Ag-Pd alloys are becoming important as ceramic and crown and bridge alloys for dental restorative applications. More recently, some commercial alloys in the market containing significant percentages of In are being promoted as tarnish and corrosion free for oral applications. The objective of this investigation was to evaluate the corrosion and tarnish behaviour of selected Ag-In and Ag-Pd-In alloys. The corrosion and tarnish behaviour were evaluated by potentiodynamic anodic polarization in 1% NaCl solution at a pH of 7.0 and by alternate immersion and wiping in 0.5% Na_2S solution at a pH of 12.0 respectively.

MATERIALS AND METHODS

Two groups of alloys were prepared by induction melting. Tables I and II list the nominal compositions of the two groups of alloys respectively. Group I typically represents binary Ag-In alloys with 0, 5, 10, 15 and 20% In in alloy numbers 1, 2, 3, 4 and 5 respectively. Group II alloys represent a quasi binary system with Ag and Pd contents varying while In content is kept constant. All the listed compositions are by weight percent. The alloys were rolled and cut to $1cm^2$ area. All the specimens were subjected to an identical high temperature treatment at 650 degrees C for 30 minutes to obtain a homogeneous alloy structure, free of possible segregation effects. The specimens were mounted metallographically. The corrosion specimens were subjected to a uniform metallographic polishing treatment through 600 grit emery paper. The specimens for tarnish analysis were metallographically polished as above and in addition were also polished with 0.3μ alumina.

TABLE I   Nominal Compositions of Group I Alloys

Alloy #	% Ag	% In	Type
1	100	-	Experimental
2	95	5	Experimental
3	90	10	Experimental
4	85	15	Experimental
5	80	20	Experimental

TABLE II   Nominal Compositions of Group II Alloys

Alloy #	% Ag	% In	% Pd	Cu+Zn	Type
6	85	15	-	-	Experimental
7	70	15	15	-	Experimental
8	55	15	30	-	Experimental
9	45	15	40	-	Experimental
10	68	17	10	5	Commercial

The corrosion behaviour was evaluated through the standard anodic potentiodynamic polarization in 1% NaCl solution at a pH of 7.0. Polarization was performed in a commercial unit (Princeton Applied Research Lab corrosion test set-up model #331-2). The unit contains a potentiostat, a universal programmer, a logrithmic converter, an X-Y recorder, and a corrosion cell. Saturated calomel electrode was used as the reference electrode. The polarization test was done in stagnant solutions with no deaeration. The potential scanning rate used was 0.1 mV/sec.

The tarnish set-up used was the alternate immersion and wiping test apparatus. A description of the set-up is given elsewhere (Tuccillo and Nielson, 1971). 0.5% Na S solution was used as the tarnish medium. The pH of the solution was 12.0.

EXPERIMENTAL RESULTS

Table III shows the initial and half hour rest potential of pure Ag and Ag-In alloys. It appears that no significant shift of rest potential values either with composition or with time during the initial half hour is detectable. This would indicate that there is little tendency for spontaneous passivation of the alloys in the chloride medium.

TABLE III   Rest Potential Values of Ag-In Alloys

Alloy	OCP Values Initial	Half Hour
Pure Ag	-10	-22
Ag- 5% In	0	0
Ag-10% In	-22	0
Ag-15% In	-17	-11
Ag-20% In	-80	-107

Figure 1 shows the anodic potentiodynamic polarization profiles of pure Ag, Ag-5% In, Ag-10% In, Ag-15% In, and Ag-20% In alloys. The polarization profiles of all the alloys are characterized by sharp increase of current density, a low anodic overpotentials above the rest potential. For pure Ag, the current density increase can easily be attributed to the formation of an AgCl film. The reaction leading to the formation of AgCl film is given by:

$$Ag + Cl - AgCl + e$$

The observed overpotential at which the sharp current density increase occurs is approximately +20 mV (SCE) which closely corresponds to the equilibrium electrode potential of -222mV (SHE) reported for the above reaction (Latimer, 1953). Polarization profiles of Ag-5% In, Ag-10% In, Ag-15% In and Ag-20% In alloys also indicate that the primary anodic process in all of the above alloys is identical in that a sharp current density increase occurs at a potential close to the equilibrium Ag/AgCl electrode potential. Thus, chloride corrosion of Ag is not inhibited or significantly enhanced by the presence of any In in the form of a solid solution up to 20%.

Table IV and fig. 2 shows the rest potential values and anodic potentiodynamic polarization profiles of Ag-15 In, Ag-15 In - 15 Pd, Ag-15 In - 30 Pd and Ag-15 In - 40 Pd alloys together with a commercial alloy. It is seen that the rest potentials show a tendency to shift in the noble direction during the initial half hour when the alloy contains Pd. More important, however, are the observed changes in the anodic potentiodynamic polarization profiles of alloys containing increased Pd. In particular, the anodic overpotential at which significant current density increase occurs shifts in the noble direction with an increasing Pd content. It would appear, therefore, that Pd alloying introduces an over voltage to the Ag/AgCl reaction and thus sharp current density increases occur only at relatively higher anodic overpotentials above their rest potentials in the alloys containing higher Pd content. In addition, passivation effects are observed at overpotentials of > +500mV (SCE).

Tarnish using alternate immersion and wiping technique in 0.5% Na_2S solution indicated that the Ag-In alloys tarnished in the order Ag>Ag-5 In>Ag-10 In>Ag-15 In>Ag-20 In. However, tarnish was significant and detectable both macroscopically and microstrucurally in all alloys. In the alloys containing Pd, there was no detectable tarnish at all on the Ag-15 In - 40 Pd alloy. Both the Ag-15 In - 15 Pd and Ag-15 In - 30 Pd alloys showed an insignificant but detectable surface discoloration. The commercial alloy also showed some tarnish but not significant.

T.K. Vaidyanathan & A. Prasad

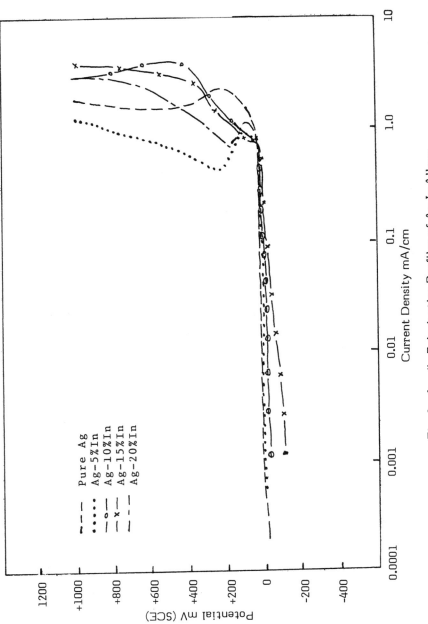

Fig. 1 Anodic Polarization Profiles of Ag-In Alloys

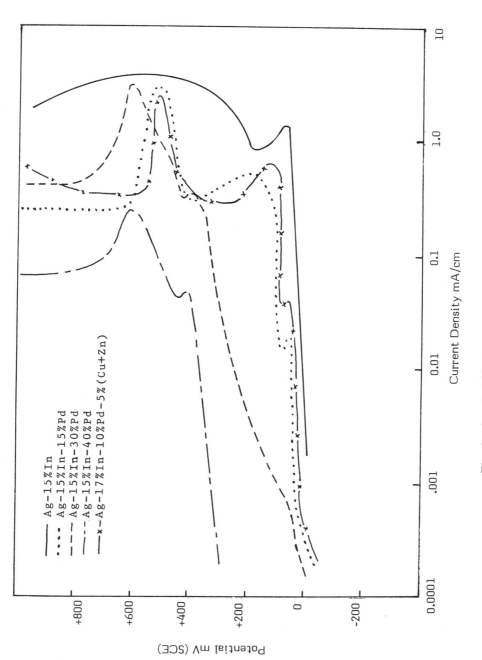

Fig. 2 Anodic Polarization Profiles of Ag-In-Pd Alloys

TABLE IV   Rest Potential Values of Ag-In-Pd Alloys in 1% NaCl Solution at 7.0 pH

Alloy	Initial Rest Potential mV (SCE)	Half Hour Rest Potential mv (SCE)
Ag-15% In	-17	-11
Ag-15In-15Pd	-40	-90
Ag-15In-30Pd	-100	-66
Ag-15In-40Pd	+28	+43
Commercial Alloy	-18	-94

DISCUSSION OF RESULTS

The absence of any improvement in the chloride corrosion resistance of Ag by alloying with In is understandable. In is a nontransition element and is also electrochemically less noble than Ag. Ag being a nontransitional element shows little tendency for chemisorption of oxygen in aqueous media. The sharp current density increase of pure Ag at low anodic overpotentials above the rest potential is a result of the low polarization effects associated with the corrosion reaction:

$$Ag + Cl^- \ AgCl + e$$

However, AgCl is insoluble in aqueous solution and therefore builds up as deposit on the electrode surface. The AgCl film is not very protective because of its porous nature. (Vermileya, 1963). However, a limiting current density at moderate to large anodic overpotentials above the rest potential is observed and this value of the limiting current density was unaltered on mechanical agitation of the electrolyte. Therefore, the limiting current density is associated with the maximum anodic processes that my occur through the film and may indicate limited protection. In may corrode in neutral aqueous media through the formation of only soluble $InCl_3$. Thus In is unlikely to alter the chloride corrosion behaviour of Ag, as observed in this study.

Pd alloying on the other hand, may increase the polarization effects on the corrosion of Ag and Ag-In alloys in two ways (Vaidyanathan and Prasad, 1979).

They are:

1. Pd, in solid solution with Ag may make Ag thermodynamically and electrochemically more noble. This would introduce an overvoltage to the Ag/AgCl reaction which now occurs at higher anodic overpotentials. For this reason, the sharp current density increase occurs at higher anodic overpotentials with increasing Pd in the alloy. Observe that the potentials at which sharp current density increase occurs in group II alloys in fig. 2 shifts from 0mV (SCE) for the Ag-15 In alloy to +300mV (SCE) for Ag-15 In-40 Pd alloy with intermediate values for alloys containing intermediate Pd contents.

2. Interestingly, passivation effects are observed at higher anodic overpotentials of ⊁+500mV (SCE) in alloys containing Pd. This can be easily explained by the fact that Pd is a transition noble metal with a 0.6 electron vacancy in the 4d band per atom. Consequently, Pd can be expected to show a tendency for chemisorption of oxygen at higher anodic overpotentials. It is now generally accepted that chemisorption of oxygen may lead to the formation of oxide films through nucleation and growth. In the case of Pd, this may result in Pd O formations (Vaidyanathan and Prasad, 1979; Rao, Dajanovic and Bockris, 1963); thus the passivation effects observed in Ag-In-Pd alloys can be attributed to the chemisorption and oxide film formation.

The tarnish resistance increase is observed in the Ag-In-Pd alloys with increasing Pd content may also be attributed to the increasing thermodynamic nobility of alloys containing higher Pd.

CONCLUSIONS

1. In alloying does not increase chloride corrosion resistance of Ag. The effect of In alloying in improving tarnish resistance is detectable, but significant tarnish is still observed in Ag-In alloys containing up to 20% In.

2. Pd enhances the corrosion and tarnish resistance of Ag and Ag-In alloys.

REFERENCES

Latimer, W.M. (1953). "Oxidation Potentials", Prentice Hall, New York.

Rao, M.L.D., Dajonovic, A., Bockris, J.O.M. (1963). "Oxygen Absorption Related to the Unpaired d-electrons in Transition Metals", J. Phys. Chem., 67, 2508.

Tuccillo, J.J. and Nielsen, J.P. (1971) "Observations of Onset of Sulfide Tarnish on Gold Based Alloys, J. Prosthet. Dent., 25, 629.

Vaidyanathan, T.K. and Prasad, A. (1979), "In Vitro Corrosion and Tarnish Characterization of the Ag-Pd Binery System", In Press J. Dent. Res.

Vermileya, D.A. (1963), "anodic Films", In: Advances in Electrochemistry and Electrochemical Engineering, Volume II, Interscience, New York (1963).

Security

INTERNATIONAL PRECIOUS METAL CONFERENCE

SESSION N -- SECURITY

TOPIC: INTERNAL CONTROL AND SECURITY

Frederick F. Schauder, Engelhard Industries

PREMISE

The risk of loss of precious metals is greatest from the individual
or collusive efforts of employees/customers/vendors. While real, I
believe, the threat of violent theft, at a facility such as Engelhard, is remote.

SOLUTION

The best security over precious metal inventories consists of a
strong, well managed system of inventory and accounting controls.

ENGELHARD INDUSTRIES SOLUTION

An integration of the efforts of all personnel involved with safeguarding over our assets. Sharing of knowledge on risk and controls.
Constant, open discussion between plant, accounting and security
management level personnel.

(a) Policy

   - Security/control policy reviewed, approved and supported
      by highest levels of management.

   - Set minimum standards to which all plants must comply.
      Determined based on experience, knowledge of operations
      and with involvement of all key personnel.

   - Monitored on day-to-day basis by plant accounting staff,
      on less frequent but repetitive basis by company level
      accounting and control experts and at least annually by
      an audited inventory.

(b) <u>Basic Policy</u> - Provides for:

1. Responsibility
2. Accountability
3. Physical security
4. Physical verification

(c) <u>Responsibility</u> - is a system of assigning physical responsibility for precious metals first to a plant and then within
the plant to the individuals and work areas where the metal
is located. Always vests with plant management but is specific
to the individual who is working on it. Never allows for joint
responsibility or allows for metal <u>not</u> to be charged to an individual.

Metal moved is verified at each step as to weights and identification. Both parties must agree or the movement is not made.

(d) <u>Accountability</u> - is the paperwork confirmation of physical
responsibility discussed above. First paper generated on
receipt of material and follows metal as it moves through the
plant until it is shipped. Each individual must initial this
paperwork to accept and pass on responsibility for the precious
metal. Copies of all paperwork provided to accounting which
maintains the responsibility ledgers. Range from simple hand
posted systems to sophisticated on line computer systems.
Accounting should <u>not</u> have physical responsibility for the
metal.

(e) <u>Physical Security</u> - includes all aspects of plant security,
against theft or other occurrence such as fire. Covers
plant construction, perimeter fencing, vaults, mechanical
aids such as television scanning in and out of the plant,
guard force composition and responsibilities, plant ingress
and egress, etc. Also includes alarm systems and monitors
and relationship with police and fire authorities.

(f) <u>Physical verification</u> - consists of two areas:

1. Operations - the daily balancing and accounting for metal
   issued to manufacturing against production, scrap, shipments and material returned to vaults. The most important
   control.

2. Accounting - Ranges from weekly to annual physical inventories to verify accounting records and physical existence
   of metal.

Security is an integral part of the internal control system. It
should participate in developing controls and be constantly knowledgeable of control systems and personnel. It should be kept
informed on the efficacy of the systems and by observation on the
attitude and dedication of the personnel charged with the management
of these systems.

Meeting the Challenge of the Rising Crimes Against Precious Metal Facilities

T. L. Weber

Thad L. Weber and Associates
Suite 825, 1055 E. Tropicana Blvd.,
Las Vegas, Nevada   89109

ABSTRACT

Security is a fluid science requiring continual adjustment and strengthening to deter criminals practicing new tricks and skills. Today we discuss these needs in relation to the major upheavals in silver, gold and other precious metal values.

KEYWORDS

Precious Metal Values and the Media Spotlight; the old gold and silver dealer explosion; high line security; hit and run burglary and robbery attacks.

INTRODUCTION

While the need to improve security in precious metal operations is a continuing process this discussion focuses on unusual conditions created by the extraordinary rise in silver and gold prices. The silver user must now rapidly develop a security program equal to that which a gold user established to protect $35 per ounce gold, while the latter must now deal with his vulnerability to huge crime losses by hit and run burglars, robbers and thieves who are attracted to the $500 per ounce gold market.

WHEN SILVER TREADS AMONG THE GOLD - IT'S A BRAND NEW SECURITY BALL
GAME.

With the unprecedented increase in precious metal values during the last half of
1979, most recyclers and fabricators of precious metals quickly learned the need
for drastic changes in materials handling and related security.

The media spotlight on the new monetary values of precious metals was not lost on
the professional criminal, amateurs or many individuals employed in the precious
metal industry.

And as countless thousands of homeowners searched their attics and their dining
rooms for old gold and silver, so did burglars who found that ransacking most
suburban residences produced a Comstock-like bonanza. Ironically, both owner and
thief soon discovered there are many new outlets offering to purchase precious
metals at relatively high value with no questions asked as to rights of ownership.
Indeed, a seller need no longer fear the IRS or the Fagin-like influence of the
traditional fence. Of course a few thieves discovered that Aunt Sara's marvelous
wedding present of 20 years ago was actually silver plated or gold filled.

Inevitably, some in the industry also succumbed to gold or silver fever. Retail
jewelers quickly became old-gold dealers. In some cases, unsold finished goods
lines were sold to refiners to take a quick profit. Indian jewelry dealers
grasped the opportunity to melt down unsold inventory and to recoup losses suffered in the decline of this market in recent years. Indeed, much heirloom silver
and valuable coins are now lost forever in the furnaces of recyclers throughout
these lands and overseas.

As gold reached the $500.00 mark and silver soared into the $30's recyclers and
precious metals fabricators, as well as smelters discovered they really didn't
know all the new crime score and its impact on profits and insurability.

For example, silver users who heretofore were able to maintain fairly cost
effective security programs without the investments in vaults and sophisticated
alarm systems learned that the value per ounce factor no longer afforded them a
margin of security as compared to gold values.

Indeed, hit and run burglars aided in some cases by ineffective security forces
were able to make away with tons of coin, sterling and fine silver in both work
in-process and scrap forms. In one instance, a crime with a loss reputed to exceed
$400,000 was successfully executed from an open stock area literally under the
eyes of security personnel.

Other silver users have reported organized and elusive efforts by groups of employees who have successfully raided the coffers and literally walked out the door
with the silver in the absence of any metal detection or package inspection
program.

Clearly the meteoric rise of silver from $6.00 to $36.00 per ounce in less than
a year dramatically illustrated a principle of security which cannot be overlooked
or treated speculatively. That is to say, security is an art as well as a science.
The individual responsible for safeguarding precious metals must constantly study
the skills and the aggressive posture of the criminal in relation to his psych and
then apply the correct technology to improve upon company security policies and
related security systems if he is to deter a moderately intelligent criminal
effort. In the case of silver, painful lessons taught us that security programs
which were adequate at $6.00 per ounce were woefully lacking when $36.00 incentives
turned many heads.

WHAT TO DO?

The silver refiner-supplier-fabricator must now consider firm action including:

High Value Areas

The Construction of bank-type vaults equipped with burglary resistant vault doors, time locks and day gates for the storage of the bulk of the high-value raw material and clean scrap. These vaults must be equipped with sophisticated and improved burglar alarm systems which incorporate "high-line" security devices which afford a high degree of resistance to compromise attack by skilled electronic burglars.

In some instances, a risk may consider the use of burglary resistant safes as an alternative, but generally the bulk requirements of silver and the difficulty in manhandling silver bullion, billets, plate, etc. will generally preclude the use of a safe for efficient storage. This utter necessity for strong physical barriers and sophisticated alarm systems will require changes in the silver processor's traditional methods of operation to provide for closed period storage of work in-process materials which in the past were left lying in place in the factory or refinery area at the close of the work day.

Work in Process Areas

Nevertheless, the silversmith, refiner or fabricator will find that despite all reasonable efforts the amount of precious metal values left out of the vault at night still add up to a tidy sum to attract many a criminal. Accordingly, it will be necessary to reassess the plant area to determine the vulnerability to theft and burglary which exists at the close of business in various high-value pockets within the premises. Wherever possible, these accumulations should be protected by the installation of delay barriers, such as cages, gates, grill-work and strong-boxes or lockers.

Alarm Systems

Management must now make certain an elaborate premises burglar alarm includes contacts on all perimeter openings adjacent to these areas, and motion sensing devices which will detect the entry of the intruder at the outset of his attack, and further that such devices are properly maintained and regularly tested. Walk testing of motion sensing devices is of the utmost importance, since such devices are not fail-safe and experience teaches us that in a factory environment declines in sensitivity, while common, are seldom discovered until the criminal exploits the weakness of the motion detector system.

Weaknesses in Proprietary Security

The effectiveness of any round-the-clock guard force, or watch tour-maintenance "on site" personnel counted on to provide security against burglary and robbery must be seriously re-evaluated. Alarm systems used to protect vaults, safes and cages or high-value storage areas must be supervised only by competent, remote central station or police stations with personnel who can supervise alarm system openings and closings and respond effectively to alarm or trouble signals. No longer should one rely on alarm systems which can be controlled at the guard or maintenance station by shunt key or deficient procedures, nor should company guards or maintenance personnel ever be permitted to respond alone to alarm signals emanating from a high-value area. Indeed, the recent increase in compromise burglaries and defeats of alarm telephone circuits by other means, such as cutting the alarm and telephone cables adjacent to the premises, or near the central sta- tion, teaches us that effective response requires that both police and management supervisory personnel must promptly proceed to the premises and enter, search and

resecure the areas from which alarm signals were transmitted. Further, in those
instances where damage to telephone cables or alarm equipment prevents the restoration of the alarm, security personnel and supervisors must remain at the premises.
to protect the assets.

Security Against Robbery

In the past, the value of silver per ounce worked against the value to be gained
in an armed robbery attack. This is no longer true. In 30 minutes or so, a band
of robbers can, using company material handling equipment, move a substantial
amount of material out of the plant into your or their vehicles. Thus security
management must now rethink its operating vulnerability as it relates to criminals
with masks and shotguns who might score well within a short time frame by invasion
of the plant during periods when limited personnel are at hand (and burglar alarm
systems are turned off). Rigid procedures requiring vault storage areas be
physically secured and alarmed prior to or coincidental with the departure of the
main work force are essential. Further, these vaults should be equipped with
time locks which will deny anyone the ability to re-enter the vault during non-
regular business hours. Appropriate decals should be used to advertise this
deterent capability. Similarly, procedures governing the operation of alarm
systems and the programming of time locks should prevent any high-value vault
storage area from being accessed in the morning prior to the period immediately
preceding the arrival of the full work force. It is to be noted that these common
sense procedures also offer a strong hedge against a planned, kidnap robbery
hostage attack which are now occurring more frequently against precious metal and
jewelry industry principals.

Theft Prevention

Of equal importance, the silver processor must devote more effort to the improvement of security against theft. From experience we learn that most theft losses
committed by employees or outsiders are seldom identified as thefts, nor is sufficient evidence developed to qualify such losses under crime or fidelity insurance
coverage. In other words, theft losses are on you, and if undetected for long
periods may threaten the profitability - the very existence of the enterprise.
Analysis of actual crimes confirms that once-thieves develop successful techniques,
they continue to practice these thefts to avail themselves of steady, long term,
non-taxable additional income.

Today the silver user is more vulnerable to such crime losses. Yesterday's silver
values tended to inhibit or limit theft dollar levels which could be absorbed
by a company within the framework of such tolerable descriptions as "shrink,
work-loss, or mysterious disappearance" without causing a business heart failure.
But this is no longer the case. In fact, the potential losses are now complicated
and multiplied by managements inability to manage inventories developed through
expanded refining activities and widely fluctuating day-to-day metal prices. Anyone who doubts such vulnerability need only reflect on the timeliness and accuracy
of their inventory control procedures and physical inventory programs. For example,
experienced gold users usually inventoried their raw materials monthly and did so
even when gold was at $35.00 an ounce. Most silver processors on the other hand
still limit their efforts to an annual physical inventory. With silver at $16.00
or $32.00, should not silver physical inventories be performed monthly or at least
quarterly?

In the same vein, silver users believed it unnecessary to install elaborate metal
and package inspection systems or to provide adequate access control systems to
restrict the removal of precious metals from high-value storage and work in-process
areas. Nor have they resorted to applicant screening and background investigation
techniques to reduce their vulnerability to the employment of individuals with

little or no integrity, (and in some cases, with extensive criminal records).

With regard to theft prevention, silver processors can reflect on both good and bad news. On the positive side, silver prices for the short and long terms will probably still mean that available crime prevention technology will be effective in deterring or detecting individuals attempting to remove moderately small amounts of silver from security areas. Electronic metal inspection systems capable of detecting a few penny-weights are available and when complemented by properly trained, diligent security personnel who will conduct an effective search of lunch pails, purses and wallets, equate to real theft prevention. However, similar efforts must be devoted to physically controlling access in and out of the security area to prevent individuals from by-passing the metal inspection system. This phase must include day annunciator (24 hour alarm devices on all exterior openings not authorized for exit except during emergency) and for adequate control or surveillance of other perimeter openings such as shipping dock doors by closed circuit TV surveillance, electrically controlled entry-exit foyers and by procedures which restrict the "access" of such areas by company employees and outsiders including delivery and pick-up vehicle drivers.

Further, it is most important to maintain these access control and inspection procedures during every minute the plant is open to any employee or supervisor.

The problems inherent in access security systems in silver processing and fabrication facilities relate to the size of the work force (usually much larger than in a refining operation), any multiplicity of shifts, and the cost of properly implementing metal inspection and access control programs with relation to the size of the work force and the hours worked. For example, for metal inspection to be fully effective it requires ten to thirty seconds to inspect the individual and his personal effects each time that person departs the security area. When you multiply this by the number of employees it is easy to visualize the traffic jams at break times, luncheon and end of shift periods unless efficient programs are developed. One approach may include a random inspection program which puts both the company and the employee in somewhat of a Russian roulette scenerio.

A Case for Integrity Testing

Thus one can readily appreciate the interplay of integrity testing and effective background screening programs for both applicants and existing employees as supplementary or alternative theft prevention techniques. Unfortunately, the larger the size of the work force, the greater the difficulty in implementing polygraph, written integrity testing or effective background investigation programs. Nevertheless, the security against theft program demands the full attention of management and labor relation supervisors to develop an effective deterrent for the future.

Transit Security

The security of silver in-transit requires an in depth re-evaluation of security programs to meet current needs. Numerous reports of truck shipments hi-jacked, thefts from airport security areas and from pick-up and delivery vehicles making daily routes demand reappraisal of your vehicle security programs to include, where possible, two-way radio communication between drivers and a manned security center, the use of strong boxes attached to the frame of the vehicle, burglar alarms and adequate training and screening of driver personnel.

Crime Insurance

As silver prices sky-rocketed, many companies suddenly discovered they were underinsured insofar as plant inventory and transient coverage was required, and were unable to obtain additional coverage without paying a significantly higher

premium rate, and/or under the conditions of making substantial improvements
in existing programs. This situation was aggravated by the sharp increase in
silver industry crimes and by the insurance industries fear of the publicity
given precious metal values. In many cases, the insurers emotions have salted
the premium waters. Nevertheless, one must reconsider the extent of insurance
limits to satisfy this question: "does your crime coverage equal your potential
catastrophe losses?" Anything less is not only dangerous speculation, but one
which exposes one's flank to officer and director liability problems.

AND NOW THE GLITTER OF GOLD

If you wonder why I have addressed my remarks primarily to the silver market, it
is simply a matter of experience in the industry and a need to establish the
priority! We recognize the level of security in force in silver processing
plants is now oftentimes inadequate to relate to today's silver values and criminal skills. On the other hand, the increase in the value of gold, however
dramatic, reflects a base of $200.00 gold which has existed for two years, during
which time the natural process of criminal attacks caused most gold processors to
upgrade security as related to burglary, robbery, kidnap-hostage, theft and hi-
jack exposures.

Nevertheless, attacks against gold users in recent months clearly indicate new
criminal tricks will continue to be directed toward gold stocks. In meeting this
challenge, the gold user-processor must now "fine tune" the security programs in
effect.

The Golden Age of Alarm Systems

Insofar as burglar alarms are concerned, there is a necessity for consideration
of the use of two alarm systems separately and remotely monitored to reduce the
vulnerability related to the electronic compromise or defeat of a single system
or the loss which may be attributed to human error such as a failure to respond
properly to alarm or trouble signals at either the alarm private central station
or a police station monitoring facility.

In addition line security systems protecting the vital telephone link between
a safe, vault, or stockroom alarm control instrument and the central station or
a police station must be upgraded to a pseudo random high line security level
which meets today's criminal (and "U.L.") tests. Tone, interlok, reverse
polarity and other lower levels of line security must now be considered passe.

And Then There's Robbery

Robbery is a highly charged emotional **moment** for both robber and victim. Best
understood the robber starts "high" and his tension increases in direct relation
to the time frame required to complete the attack (e.g. how much gold can the
robber get in the least amount of time he must sustain a high level of stress
while holding company employees and others under duress). Thus when gold reaches
a higher value, the robber dreams of high stakes in an attack which takes only
seconds or minutes, and so as the prices increase so do the odds for robbery.
Accordingly, it is imperative that defenses against robbery be re-evaluated to
make certain safes or vaults used for high-value gold storage are always under
time lock and alarm supervision at any time when a full work force is not at
hand. The safety in numbers concept is based on actual experience. There have
been few, if any, robbery attacks against gold processing facilities when there
were 25 or more employees effectively at-hand in an area where the robbery attack
would have been focused. Processors whose work forces are smaller in numbers
must consider other security measures such as physical access barriers, the use
of off-duty law enforcement officers or private security forces.

The Midas Touch

When it comes to theft prevention, gold users now face the more complicated problems. Simply stated the capability to detect a penny-weight or two of gold surreptitiously removed through a metal inspection and/or package inspection process is one exceedingly difficult to accomplish. Previously established metal inspection programs must now be re-evaluated to establish a cleaner environment, one free of any interference from moving metal, electrical energy fields, ultrasonic cleaners, induction furnaces, time clocks, etc. e.g. the metal detection system must be able to distinguish a few pennyweight moving through the detector in the pocket, mouth or other place of personal concealment. It must be able to distinguish this metal from any extraneous force.

In addition the employee must be taught to "cooperate" by leaving any personal jewelry or other metallic objects at home or in a secured locker outside the defined high security work area.

Security guards assigned to perform metal inspection must also be selected for integrity as well as diligence and tact since it will be their responsibility to maintain employee cooperation and goodwill even when at times they must "scan" the individuals by hand to determine a metal source. Indeed, in practical experience there is clearly a need for management to maintain monitored and video recorded closed circuit TV surveillance of metal detection stations in order to evaluate the true effect of the program. We note too that female security officers are potentially more effective in such an assignment.

Metal Inspection without exit control = Zero

A high percentage of the work force earning a fair wage and working on a controlled security environment are unlikely to resort to theft. But to the hard core thief now on your payroll (or seeking employment with you) theft is profitable, penalties for apprehension are not meaningful, and he considers metal inspection a "challenge". Give this individual an unsecured window, a fire exit without a day alarm device, a mediocre metal inspection process, or any other loophole and he will take immediate advantage of the security faux pas - chances are these criminals already have some metal stashed away under machines, benches, in duct work, lockers, etc., just waiting for you to let down your guard.

A Defined Integrity Testing Program

Given the limitations of the metal inspection process, and the difficulty in detecting gold in salt, solution, powder or dust form, there is a strong argument for other deterrents to theft by employees. Some companies have filled this need with a random selection polygraph program administered with the voluntary support of the work force.

The essential conditions include: * the concept of "amnesty" - management concerns itself only with deterring future theft, therefore an individual need only deal with the question of the theft of company property which took place after the amnesty polygraph program was established.

* The element of random testing. A qualified polygraph examiner should test a pre-determined percentage of the work force each month. However, supervisory and hourly employees tested should be selected on a random basis, wherein any employee must contemplate the possibility of such a test within a thirty-day period.

To be sure, such a program requires careful consideration in terms of legal and

ethical standards. It may also require modification to suit labor relations and
the costs of such programs. Nevertheless, many employers have succeeded in estab-
lishing the program with the formal or tacit approval of unions, and in obtaining
the voluntary participation of personnel in critical security assignment.

Maintenance of this program coupled with a policy covering pre-employment polygraph will produce long term results even if 100% voluntary participation cannot
be achieved initially on the part of the existing work force.

Recent Trends

Plating and stripping operations are no longer immune to crime loss. Today there
are frequent reports of burglars pumping out gold solution into a vehicle parked
adjacent to a plating area, anodes fished out of silver tanks, hijacking of
trucks carrying concentrated slimes destined for refining, and these criminals
do not seem to have any problem disposing of the materials in this form. These
events spell much trouble, for plating operations have traditionally lacked both
physical and electronic security. Further, considering the nature of these
facilities physical security improvement is usually not practical while the
corrosive environment challenges the alarm installations to provide a system
which will be both effective in detecting intrusion and relatively free of equipment failure.

Future Security

For the long term, it is essential for precious metal companies to strengthen
security in existing facilities.

1. The common use of two or more remote monitoring facilities providing security
against burglary of any significant precious metal operation.

2. The more extensive use of motion sensing equipment to provide almost complete
coverage of the interior of any such high security facility. In this connection,
it is noted that the expansion of motion sensing devices in refining, manufacturing and smelting environments is loaded with potential false alarm problems.
This expanded security need will then run in an opposite direction to law
enforcement efforts to both regulate and reduce unnecessary alarm response by
fines, discontinuation of police response, etc. The alarm system user must
re-evaluate the capabilities of his alarm contractor and in some cases clean up
the plant environment to reduce false alarms. This is also important in maintaining the more effective response attitudes on the part of the police, alarm company
guards and company supervisors, since excessive false alarms soon create a "cry-
wolf" response attitude that leads to significant loss.

3. The extended use of closed circuit TV surveillance equipment including video
switching and video tape recording throughout the business day (in some cases,
during the closed periods as well). Today's management supervisors spend more
time away from the direct supervision of the business, and therefore should rely
more on closed circuit TV recordings as a supplement or backup to their "presence"
in the shop.

4. Unfortunately, no significant technological improvement in metal inspection
equipment is anticipated. However, it will be necessary to make modifications in
plant layouts and in inspection techniques so as to put teeth in the metal and
package inspection programs.

5. Finally, management must strive to develop a work force with a higher
integrity level to combat the difficult social and economic pressures which exist
at this time. Begin now to utilize written integrity testing techniques, to

reconsider the use of polygraph, and to recognize employee integrity (or the lack of it) in terms of work habits such as tardiness, absenteeism and a proclivity for workman's compensation and unemployment claims. It is also more likely an applicant of lessor integrity standards will accept an employment offer requiring extensive training when in fact, the individual has no long term interest in the job.

Planning Ahead

Security in design should be the first priority of any new precious metal facility. Elimination of windows and other extraneous openings, single points of employee entry-exit, enclosed shipping and receiving areas, secure entry foyers, fenced employee parking (segregated from the security building), bank type vault construction, adequate lighting and burglary resistant locks and hardware are basic considerations in planning a secure operation.

Similarly, if you plan to continue operation in an existing facility,a capital improvement program to achieve these physical security standards will pay off in terms of prevention of crime losses and reduce security labor costs.